DEVELOPMENTAL AND CELL BIOLOGY SERIES

EDITORS
P. W. BARLOW   P. B. GREEN   C. C. WYLIE

DEVELOPMENTAL
BIOLOGY OF *PHYSARUM*

DEVELOPMENTAL BIOLOGY OF *PHYSARUM*

HELMUT W. SAUER

Department of Biology, Texas A & M University

CAMBRIDGE UNIVERSITY PRESS

CAMBRIDGE

LONDON   NEW YORK   NEW ROCHELLE

MELBOURNE   SYDNEY

Published by the Press Syndicate of the University of Cambridge
The Pitt Building, Trumpington Street, Cambridge CB2 1RP
32 East 57th Street, New York, NY 10022, USA
296 Beaconsfield Parade, Middle Park, Melbourne 3206, Australia

© Cambridge University Press 1982

First published 1982

Printed in Great Britain at the
University Press, Cambridge

Library of Congress catalogue card number: 81–21682

British Library Cataloguing in Publication Data
Sauer, Helmut W.
Developmental biology of Physarum
—(Developmental and cell biology series; 11)
1. Physarum polycephalum   2. Fungi—Growth
I. Title   II. Series
589.2'9   QK635.P5
ISBN 0 521 22703 8

Contents

Preface

The purpose of this book is to organise facts and fancies about the life cycle of *Physarum*, a true slime mould and a remarkable organism, in an effort to understand how it develops from a microscopic cell to a unique macroscopic creature.

K. E. Wohlfarth-Bottermann, the eminent cell biologist and authority on shuttle streaming in *Physarum*, has been the catalyst to get this project started; he suggested, after the author gave a seminar in his institute at Bonn University in the autumn of 1977, that the developmental biology of *Physarum* would deserve a place in the Developmental and Cell Biology Series of the Cambridge University Press. Here A. Winter deserves credit for his interest and patience and A. J. Colborne for editing the manuscript.

Many reviews of various aspects of research on *Physarum* have appeared and an excellent monograph has just been published containing six essays that attempt to summarise the current 'status of experimental analysis on the problems of growth and differentiation presented by *Physarum*' as stated by the editors W. F. Dove & H. P. Rusch (1980). In addition, a comprehensive treatise with articles by some 25 authors, covering every aspect of the cell biology of *Physarum* in great detail, will soon appear in print (edited by H. C. Aldrich & J. W. Daniel).

Some parts of this book overlap with the other publications, but as a whole it is intended to give a coherent view of *Physarum* as a developmental system. At the same time, it is also the record of how a curious individual, with a background in classical embryology obtained as a disciple of F. Seidel at Marburg University, has become convinced, after having held a critical attitude towards so-called 'model systems', that *Physarum* is perhaps the most suitable organism for a combined onslaught to solve major problems of developmental cellular and molecular biology of eukaryotic organisms.

While this view may be too optimistic, anybody with an interest in biological development must be struck not only by the opportunities offered by this humble slime mould and the amount of knowledge accumulated over the years but also by the lack of true understanding that prevails. I have tried to make students with an interest in the biosciences curious about, if not fascinated by, the lives of *Physarum* in the hope that someone may become motivated to do some ingenious experiment that will help us to unravel the mysteries of developmental biology.

I acknowledge with thanks the helpful communications with many 'Physarologists', ever since H. P. Rusch introduced me to *Physarum* in 1967 in his laboratory at Madison, Wisconsin; in particular H. C. Aldrich, V. G. Allfrey, R. Braun, O. R. Collins, J. W. Daniel, Jennifer Dee, Jessica A. Gorman, F. B. Haugli, H. R. Henney, C. E. Holt, A. Hüttermann, D. Kessler, P. Loidl, H. Matthews, Joyce Mohberg, L. Rakoczy, W. Sachsenmaier, S. Shall, W. Schiebel, Th. Schreckenbach, G. Turnock, J. Tyson, T. Ueda. Also, I note with gratitude that I have been fortunate to have had dedicated and able collaboration and many fruitful discussions over the years with: Kay L. Babcock, E. M. Goodman, Loralee Linder-Sauer, Brigitte M. Jockusch, at the McArdle Laboratory, Madison; Sigrid Esselborn, G. Wegener, at Heidelberg University; Roswita Boehme, H. H. Bohnert (Düsseldorf), G. Ernst, H. Fouquet, A. Hildebrandt, K. Scheller (Marburg), Cornelia Schicker, R. Wick, at Konstanz University; and Annete Baeckmann, G. Büchler, G. Pierron, U. Plagens, R. Simmler, Ruth Simon, at Würzburg University. I am grateful to my colleagues in the Würzburg *Physarum* group for their help in preparing this manuscript, and I am particularly indebted to Dorle Wolf for typing and preparing the manuscript for publication. My thanks are due to G. Pierron and R. Wick for careful reading of the manuscript, and to R. Wolf for his excellent photographs. Furthermore, I acknowledge the generous and continuous financial support from the Deutsche Forschungsgemeinschaft.

Last, but not least, I appreciate the perseverance and understanding of my family at times when the slime mould had gained control over our family life.

February 1981   Helmut Sauer

Introduction

Everybody will have observed, in one way or another, that any living thing changes as it attends to the necessities of life: to feed, to reproduce, and to get along with its neighbours that want to do the same.

Because changes – in shape and with time – are so obvious for all organisms, it is commonplace to talk about phenomena of biological development with no need to ask immediately 'how' and 'why'. It is self-evident that a child grows up because his parents did too, and that the embryo develops eyes because they are needed subsequently for sight. It is quite easy for us to anticipate these changes in time and space, as we know from experience that they have occurred before. We also know from mistakes, which fortunately do not often occur, that normal changes during development are useful. The unborn embryo does not know what we curious people can observe and deduce *a posteriori*, and although we can describe the developmental process meticulously, we cannot *a priori* explain how it works. The embryonic development of an individual is a uni-directional process in time: one only gets older. However, the embryologist proceeds in the other direction and looks at ever younger embryos; they get smaller in size and simpler in shape and finally reduce to a single cell.

That cell, the egg, is remarkable: it is a germ cell and once activated it develops into a specific, complex organism which produces more germ cells, eggs and sperms. The fertilised egg, as a stem cell, does not really develop just in one direction from embryo to adult, but also reverts from the macroscopic to the microscopic form of life.

While everybody knows about life cycles, how does the egg know it? Well, I do not know, and we will have to ask the right questions to find out. Today, one is tempted to say that the knowledge is all laid down in the genome, the memory store of the evolutionary process.

Philosophically and historically, two alternatives have been heatedly debated. One view, preformation, holds that the adult organism is performed in the egg, it is just much smaller; the other, epigenesis, contends that the embryo gets more complex as it develops features that were never pre-planned in the egg. Embryologists have tried to solve this problem the scientist's way: by doing experiments. However, a clear answer has not emerged as yet, although a few common principles of ontogeny have been

1

glimpsed at, and they have been labelled 'growth' and 'differentiation'. The trouble is that neither process is simple and they are by no means mutually exclusive; they both occur at the same time, often in the same place. Some other phenomena are frequently observed by the experimental embryologist. One is called 'regulation', to describe the fact that some maltreated embryos repair themselves or that, in most mysterious ways, one can become two, or many can become one individual. Comparative studies have revealed that some embryos do not 'regulate' well; they have been labelled 'mosaic eggs'. Today, it has become clear that regulative and mosaic embryos really show only one phenomenon, mostly defined as 'determination', which happens in the one type earlier in development than in the other, making timing one of the key issues of development. If there were a Nobel prize for semantics, the dozen or so terms which figuratively describe determination should be nominated, for they all point towards the central issue of developmental biology. However, it is completely unsettled, yet fascinating, that embryonic determination may well turn out to be an 'indeterminate' process. Nevertheless, there is good evidence for 'determinants' (i.e. substances, not processes) in some eggs, presumably manufactured while the eggs mature inside the mother and stored at certain places to act in a mysterious fashion later on when the embryo takes on its typical shape. Once again, we encounter a plethora of descriptive terms with the attribute 'morphogenetic', such as factors, gradients, movements, fields, even landscapes, and what we cannot explain on this basis is shoved back in time to rest comfortably during oogenesis, the most obscure period of development.

As an embryo gets a little older, and cells become arranged in blastemas, intercellular communication is required for a well organised progress of embryonic development. The term 'embryonic induction' describes this to a first approximation, and while it is not understood either, vehement debate exists as to which of two alternative hypotheses is correct. One involves an instruction process, as specific information is assumed to be passed from the inducing to the reacting blastema, teaching it what to do such that it never forgets from that moment on; the other hypothesis holds that the reacting tissue has at least two options to develop further, but needs a (unspecific) signal to select for one or the other.

Immediately, one is reminded that instruction and selection are key issues in Lamarckian and Darwinian theories, which bring up a question that has not been asked lately: are the mechanisms of ontogeny and evolution not so different after all? Of course, evolution is an important aspect of biological development, but it proceeds along such a different time scale and is associated with such massive changes in the genome that it seems far-fetched to invoke similar mechanisms for embryogenesis, which is rapid, cyclic, and supposedly involves no change in the genome.

The concepts of evolution, largely a result of comparative anatomy of adult organisms, have rested peacefully in the framework of Darwin's great theory, which has been supported by so much evidence provided by molecular biology. The recent application of Darwin's principles to sociobiology, or catchy phrases like 'selfish DNA' and the contemporary label of the 'me-generation' have dragged evolutionary theory into broad daylight. We must remember that the evolutionist records the life history, while the embryologist can at least design and do experiments on young embryos, even if he still does not know what really happens. Thus, he is in a better position than his colleague who ponders the fossil record and writes an ingenious story, or uses the computer in modelling and wonders whether fluctuations in the gene pool will suffice to explain how *Escherichia coli*, flies and elephants came into being, or why survivors survive in the struggle for survival.

However, when we consider the evolution of organisms and experiments on the development of an individual organism as two facets of developmental biology, and begin to recognise a 'comparative anatomy of development', a very surprising observation is made: those mysterious central issues, like determination, morphogenetic pattern and differentiation, are by no means restricted to embryos; they are integrated in a network of life-cycle strategies. Moreover, these phenomena are not restricted to animals and plants, but are found at all levels of biological organisation, including the cellular and molecular levels.

Just as we have witnessed how a great unifying concept of molecular biology was moulded from a multi-disciplinary approach on the right kinds of organism, i.e. bacteria and bacteriophages, so there may be simple rules behind the infinite variety of building displayed in adult organisms, as we learn to understand the developmental programme.

Today, as the replication of the ubiquitous genetic information and its transfer and function in living things have become clear enough to anticipate no great surprises or special principles of life, many bright and inquisitive people have moved to the last frontiers in biology: the brain and the egg. Shall we find our mind mirrored in developmental processes of the brain? It is too early to say, and people proceed in two directions, employing common methods of cell and molecular biology and ingenious genetic techniques. One group searches for highly specialised developmental situations that might illuminate and amplify one specific aspect of development, such as specificity of cell contact between nerve cells, or the recognition of self which involves the major histocompatibility locus – also dubbed the 'supergene'. The other group looks for the most simple systems they can find. Then they analyse the whole life cycle and try to extrapolate the findings from their model systems to other more complex ways of life. The social amoebae of *Dictyostelium* provide an excellent example of cell communication, aggregation and pattern formation. The coelenterate *Hydra* has become the model system for still-elusive morphogenetic gradients. Between the two approaches, real progress can be anticipated. Rules of development may be discovered and substituted for the fanciful stories that have long been told and a unifying, maybe even a simple, concept of development may evolve.

There have been many surprises, some in areas where we felt safest. No one predicted that the Mendel factor, or the gene, would exist as families, that structural genes would turn out to be mosaics of introns and exons, and that mRNA had to be spliced; while we used to worry about how the stable organisation of DNA can change to allow evolution through mutation, we must now wonder how the thread of life is held together, following the discovery of 'movable elements' (jumping genes), which has led to the provocative notion of a 'fluid genome'. Transcriptional control, the long-standing dogma of developmental biology that is sometimes used synony-mously with determination, may boil down to gene inactivation. And who would have dreamed that a couple of hundred structural genes are needed to build an egg shell? An embarrassing question was asked some years ago: do cells cycle? And who knows, developmental information may not reside exclusively in DNA, leaving ample scope for a constructive 'epigenesis' during the time-course of development in the absence of a rigid programme. On the whole, it turns out that to question established theories is a healthy attitude: we may find some hard facts.

Indeed, I am not aware that any organism has evolved from one kind to another under the critical eye of an observer. While it is highly unlikely for me to turn into a slime mould – although some people may begin to wonder – I have tried to imagine the problems that *Physarum* faces, when it turns from a pale minute amoeba into a bright-yellow rather large creature, which grows and grows, and in due course decides to change and make many heads (hence *P. polycephalum*) from which very many amoebae will hatch and make more slime moulds, just as they have done for a very long time.

Of course, I am prejudiced, but I maintain that this creature is in a very special position on the evolutionary tree, right where the basic principles of developmental biology of higher organisms have evolved and been kept in a 'frozen' state for us to sort out. Indeed, many pertinent phenomena of higher embryos can be conceptionally deduced from the life cycle of *Physarum*, and what is more important, the problems can be cleanly dissected and studied one by one, utilising the complete arsenal of modern biological technology, including genetics on large numbers, so far only possible with microorganisms.

I have tried to trace the life history of *Physarum* and stressed its clocks, maps and model character, in view of contemporary developmental, cell and molecular biology.

Much work has been done. I have referred to all pertinent reviews and most key research papers, but by no means to all published work on

Physarum. Much more work is necessary to answer the challenging questions that have already been raised, now that the groundwork has been laid.

Developmental events occur in time and space as continuous processes and they are hard to portray in still pictures. Therefore, I have introduced sketches, in most cases very naïve and extremely simplified, which may aid the reader to follow a sequence of developmental reactions, just as they have helped me avoid lengthy written descriptions and interpretations. In doing so, I have been amazed how difficult it is to reduce what I felt was very clear to an unambiguous picture which might allow the occasional glimpse of a basic universal developmental pathway. I have tried anyway, and the errors you will find are mine. Your correction and any comment you may have on the lives of *Physarum* will be very welcome.

1

Lives of a true slime mould

A typical slime mould exists in at least three very distinct forms of life. These are interconnected as a life cycle which has allowed for the survival of such creatures over millions of years, and for their world-wide propagation.

My favourite specimen, *Physarum*, lives in moist forests where one would expect to find fungi; its fruiting bodies, the sporangia (Fig. 1), indeed resemble those of some fungi. Therefore, it has been considered to be related to moulds. The sporangia are quite pretty, sometimes colourful, and very variable in shape, which has enabled taxonomists to distinguish over 500 species of slime mould.

Physarum always produces many fruiting bodies, hence the species name *polycephalum* which means multiheaded. Each is about 2 mm tall, fixed to its substratum and exposed to the light. Every developmental biologist would agree that the shaping of these bodies is a fine example of morphogenesis. Once created, these structures do not grow or move. They seem dead, but they are not, they are only dormant and can burst into life.

In one of its other forms, *Physarum* lives in the dark as a slimy yellow mass with no definite morphology, which must have inspired an early observer, Wallroth (1833), to name it 'slime mould' or 'myxomycete'. This creature avoids the light, hides under the bark of decaying trees, feeds on organic matter and microorganisms and grows to a considerable size, up to about 30 cm in diameter. Since it can sense a food source and crawl towards it like a gigantic amoeba, which is not typical for either fungi or plants, it has been considered to be some kind of animal, and slime moulds have been classified as the myxetozoa by de Bary (1860).

Today the place of slime moulds within biosystematics is still disputed, but the slimy matter reveals two amazing features whenever it is placed under the microscope. First, a vigorous protoplasmic streaming can be observed which keeps the contents of the whole organism well mixed. On closer inspection veins of ectoplasm can be discerned, in which the endoplasm displays a unique shuttle streaming, the flow being reversed about every minute. These regular oscillations which occur without change in the environment are a good example of an endogenous biorhythm.

The second observation is that there are millions of nuclei in the protoplasm which are not separated from each other by cell membranes. Hence

7

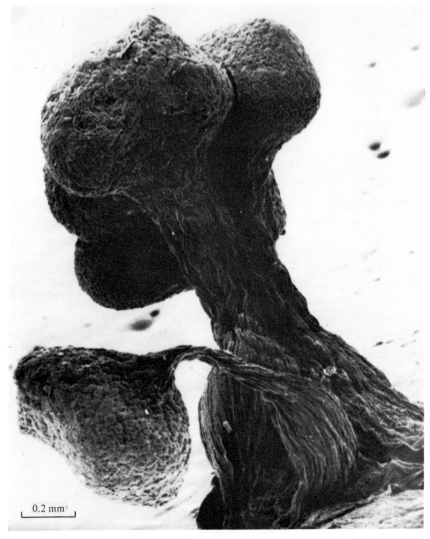

0.2 mm

Fig. 1. A fruiting body (sporangium) of *Physarum* (photo by R. Wolf, Würzburg).

this structure has been termed a plasmodium. It is much larger than a typical uninuclear cell, yet lacks specific tissues and organs found in true multicellular organisms. A most fascinating discovery was made by Howard (1932), who noticed that all the nuclei of the plasmodium divide at the same time, i.e. with natural synchrony. Later on it was found that nuclei divide after regular intervais, each of about eight hours. Consequently, the nuclear division cycle is another regular biorhythm which is maintained in the plasmodium under constant external conditions.

(a)

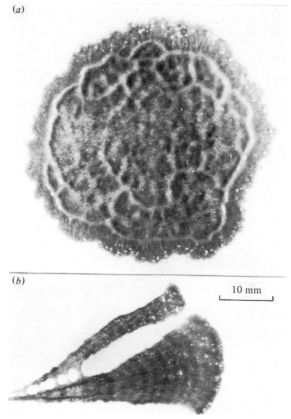

(b)

10 mm

Fig. 2. Plasmodia of *Physarum*: (a) a round flat disk of protoplasm in balanced growth; (b) a polarised individual, migrating (photos by R. Wolf, Würzburg).

Since the nuclei divide when everything in the plasmodium has doubled in mass, it gets bigger all the time. On the other hand, small pieces of a plasmodium can be cut off and each one grows again into another large plasmodium. Therefore, in this form *Physarum* can be propagated, perhaps indefinitely, by subculturing. A well-fed plasmodium (Fig. 2a) is looked on by biochemists as a giant cell. They analyse its constituents in homogenates and can be sure that the molecules they study are derived from a homogeneous source. Such a biological system is very rare, and for the developmental biologist the plasmodium is an example of pure growth, displaying the bare necessities of life, with no trace of any specialisation.

Field biologists will notice that this compact, flat, yellow disc produced in laboratory culture looks quite different compared to a plasmodium in its natural habitat. There, *Physarum* is broken up into a network of interconnected strands of protoplasm, forming a fan at the moving front

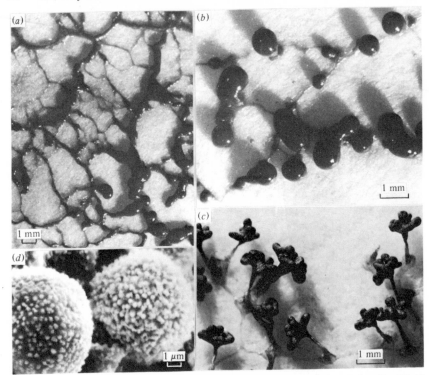

Fig. 3. Transformation of a plasmodium into sporangia (photos by R. Wolf, Würzburg). (*a*) A starved plasmodium has broken up into a network of strands; (*b*) after illumination, strands segregate into beads; (*c*) each bead forms a sporangium; (*d*) spores derived from a mature sporangium (scanning electron micrograph).

(Fig. 2*b*). They will also notice that a well-fed plasmodium does not move around at all, it is completely sedentary and just grows.

An important observation is made when a 'hungry' plasmodium rests in a room on a shelf close to the window (Fig. 3*a*). One day, after about a week, the whole thing will have become transformed into fruiting bodies (Fig. 3*c*). For the developmental biologist this is an excellent example of differentiation in that one form of life, the growing state, has completely and synchronously turned into another – an alternative state. This situation offers a very good opportunity of comparing molecules in extracts of a plasmodium and sporangia and investigating the differences between them, without being bothered about the incomplete separation of a mixed population of various cell types or the coincidence of the processes necessary to maintain two essentials of most other developmental systems: growth and differentiation.

Now we know that a plasmodium can turn into fruiting bodies, we need to know if, and how, a fruiting body can turn into a plasmodium. This leads us to the third life-form of *Physarum*, which is microscopic.

When the fruiting bodies dry out, their heads rupture and numerous small particles are released. Actually, a network of fibres, the so-called capillitium, expands explosively and disperses these particles, the spores, over some distance, thus allowing *Physarum* to enjoy a brief airborne existence. These spores are black and their walls have a distinctive appearance (Fig. 3*c*) which provides taxonomists with species-specific markers, but probably has other functions in the propagation of the many thousand progeny of one plasmodium. The cell biologist recognises that the transition from plasmodium to fruiting bodies is not only a case of morphogenesis but is also an example of the creation of cells. One may argue that nuclear division and cellular division have become uncoupled, unlike the situation in other organisms, be they unicellular or multicellular. Thus nuclear division occurs in the plasmodium and cellular division occurs later on, in fruiting-body formation. Finally, as these cells become differentiated into spores, this is an obvious example of cytodifferentiation.

Under suitable conditions, a spore can break open and release a naked cell which can multiply like any typical cell (Fig. 4), and then nuclear and cellular division occur as two strictly coupled events. In this state nothing reminds us of the macroscopic plasmodium, and everyone would classify *Physarum* as a unicellular organism, a protozoon; just one of many lower creatures that have a higher organisation than bacteria and other prokaryotes in that they possess a nucleus separating the cytoplasm from the genetic material. However, as we already know, these eukaryotic individual cells, the myxamoebae, can also exist in two macroscopic forms, one displaying pure growth and the other one differentiation.

A very important step then in the life of *Physarum* is the transition from a single cell to a multi-nucleated plasmodium. Careful analysis has revealed an amazing flexibility in just how that transition may be achieved. Developmental biologists are interested in the various possible pathways which may become models for the initiation of a life cycle in higher organism from an egg cell. Taking an optimistic view, *Physarum* may help us to understand the developmental programme that allows initiation of growth and the division of labour in more complex organisms. As we shall see below, the life of *Physarum* itself is quite complex, too, in that it not only has alternative steps available but is also a multipotent system.

For now, let us summarise the essential steps of the life cycle of *Physarum* (Fig. 5), which contains two alternative states of growth: protozoon or plasmodium, and three uni-directional and irreversible transitions: from protozoon to plasmodium (1), from plasmodium to fruiting body (2), and from fruiting body back to protozoon (3–4).

Before we begin to analyse further ramifications of the life of the true

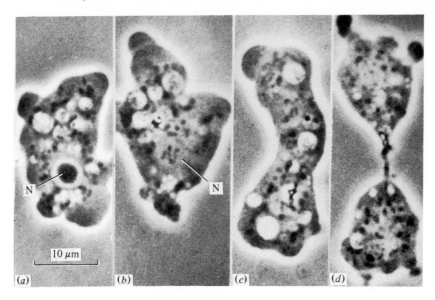

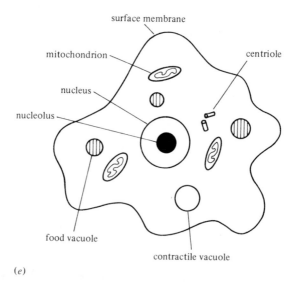

Fig. 4. An amoeba, the microscopic form of *Physarum*. (*a*) On agar surface; (*b–d*) dividing (from a film by R. Wolf, Würzburg); (*e*) a schematic sketch of an amoeba. N, nucleus.

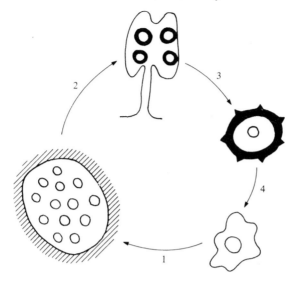

Fig. 5. The essential life cycle of the true (acellular) slime mould *Physarum*. An amoeba is transformed into a plasmodium (1) (amoeba–plasmodium transition, APT), the plasmodium can differentiate into sporangia (2) in which spores are formed (3), from which amoebae (4) emerge.

slime mould, a brief description of a 'fake' slime mould will be given, which is quite different in that it makes no slime. The organism is *Dictyostelium discoideum*, a representative of the Acrasiales, which has already established its place in cell biology (Bonner, 1967). A more appropriate term for this organism is 'cellular' slime mould which distinguishes it clearly from the myxomycetes or 'acellular' slime moulds. *D. discoideum* was discovered by Raper (1935) in forest detritus. At one stage it looks just like *Physarum*: both live as solitary amoebae, move over the same substrate with their pseudopodia; they feed and divide by binary fission. However, since neither one can be distinguished from other soil amoebae at that stage, they have no more in common with each other than with true unicellular protozoa. In all other stages of the life cycle the two slime moulds are completely different. When the food sources are exhausted, some hungry amoebae of *D. discoideum* send out an 'acrasin', a signal molecule which causes a group of about 10000 amoebae to aggregate. These amoebae become social in that they respond chemotactically to pulses of cyclic adenosine monophosphate (cAMP, one kind of acrasin) and migrate towards the leader (the centre), make contact with each other, and relay the signal by cAMP secretion to the periphery of the group. Then each little cell society becomes enclosed in an envelope that contains cellulose. It starts migrating like a naked snail about 2 mm long, which has been called appropriately a 'slug' or 'grex'. It has also been designated a pseudoplasmodium to indicate a vague similarity to the other slime moulds, but to stress the fact that each amoeba of the aggregate remains a separate entity. After a while the migration of the slug stops and a remarkable case of morphogenesis is initiated by the amoebae of the anterior portion. They become vacuolated and stretch vertically, building up a stalk which extends to about 1–2 mm, as more and more of the adjacent amoebae move in and stretch out. In the meantime the amoebae at the periphery, derived from the posterior part of the slug, encyst and finally are moved up the stalk to form the head of the fruiting body. The whole process has been described figuratively by Bonner (1967) as a fountain running backwards. This then is a structure which serves the propagation of *Dictyostelium* and survival of unfavourable times as a dormant 'spore', much like in *Physarum*. Again the fruiting bodies resemble a stage in fungal development, hence its classification as slime 'mould'. In summary, its life cycle contains four essential steps: proliferation, aggregation (socialisation), morphogenesis, and propagation (Fig. 6a). The main invention is the differentiation of amoebae into stalk cells, which will die, and into 'spores' which can survive; maybe the earliest example in living organisms of altruistic behaviour. *Dictyostelium* has long been recognised as a model system for study in developmental biology. Phenomena of cell communication, migration, specific cell adhesion via glycoproteins, cell sorting, position information, pattern formation, differentiation, and morphogenesis, along with biochemical and mutational analyses are well documented (Loomis, 1975).

So why bother to study the development of *Physarum*? Well, multicellular organisms do not develop by aggregation, but from a single cell stage, the egg, by cell division. This mitosis, however, plays no role in the development of the grex in *Dictyostelium*. Next, the decision amoebae have to make allows only one alternative: to die or to survive, whereas *Physarum* is a multipotential system. While the life cycles of *Physarum* and *Dictyostelium* look quite similar (Figs. 5, 6a), the latter can also be drawn differently (Fig. 6b). Here it is shown, as has been stressed by Garrod & Ashworth (1973), that amoebae have the option of either becoming a cyst or a stalk cell, and that cysts can revert easily to amoebae. Consequently, there may not be an irreversible commitment in the normal life of *Dictyostelium*, while in *Physarum* we have already encountered at least two alternative states: a sporangium cannot turn directly into a plasmodium, and a plasmodium cannot break up into amoebae. Therefore, *Dictyostelium* may just alternate between two reversible states, one proliferating and one not, and lack a true life cycle, being content instead with the cell cycle, as in all other protozoa. For the sake of argument we can speculate in three ways. (1) Differentiation in *Dictyostelium* more closely resembles 'sporulation' in some bacteria, and

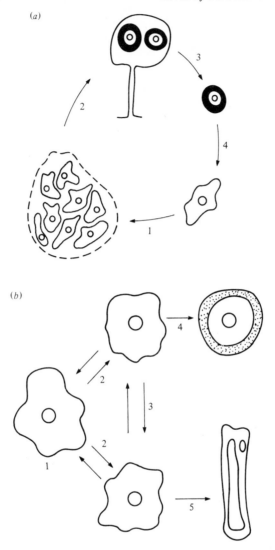

Fig. 6. The life cycle of a cellular slime mould, *Dictyostelium*. (*a*) Many amoebae form an aggregate (the slug) (1), which can differentiate into a fruiting body (2). It contains spores (3) from which amoebae emerge (4). (*b*) Amoebae have the option to either grow and multiply (1) or aggregate (2), and for some time can become either spore or stalk cell (3); but eventually they become either a spore cell (with a wall) (4), or stalk cell (with vacuole) (5).

the correlated rapid physiological changes may resemble those in pro-karyotic organisms during starvation in Nature (or step-down conditions in the laboratory). (2) The genome of *Dictyostelium* is quite small, 0.1 pg in a cell which is grown in pure culture (axenic, i.e. without other organisms), 30–40% of which is mitochondrial. It is contained in seven chromosomes in a haploid cell which can complete the 'life cycle' without a change in ploidy. Furthermore, mitosis is of the 'closed' type in *Dictyostelium* amoebae, resembling yeast cells. In *Physarum*, a typical nucleus in the diploid plasmodium contains about 1 pg DNA, and the complexity of its DNA is more than ten times that of *Dictyostelium*, and the myxamoebae divide by an 'open' mitosis, complete with centrioles and a typical spindle apparatus. (3) Like plants, which can be propagated vegetatively and, in some cases, even cloned from single cells, *Dictyostelium* may lack the mechanism of irrevers-ible commitment. That is evident in higher animals by the primary distinc-tion between germ line and the body (soma), which strictly segregates propagation of the animal species from differentiation of the respective individual's body. In *Physarum*, amoebae may function also as the germ line (in addition to their vegetative propagation) and the plasmodium as an alternative, yet large, somatic state.

Although these speculations do not explain any developmental phenomena, they may have served to demonstrate differences in cellular and acellular slime moulds. As has recently become clear, by the discovery of major differences in the structural organisation and function of the eukaryotic genome as compared to that of prokaryotes, extrapolation of results obtained from simple developmental systems to more complex ones may not always be correct. Therefore, the thorough analysis of *Physarum* may be necessary for both the understanding of its own life cycle and as a model for the development of higher organisms.

Before turning to a more detailed description of the life cycle of *Physarum*, let us stress that while it is not important whether the true slime moulds are claimed by botanists or zoologists, they occupy a major transi-tion zone from unicellular amoebo-flagellate state to three very different phylogenetic categories: fungi, plants and animals. These include very complex multicellular organisms which have become specialised with time. *Physarum* may be content to have remained unspecialised, broadly adapted to many environmental conditions and demands without any organs for separate functions, yet not without the options to develop into a bigger structure and make irreversible commitments.

Hence, the amoeba–plasmodium transition, which has not received much attention up to now, may be a footprint from which to conclude how multi-cellular organisms have evolved. In ontogenetic terms, we may learn, from this developmental transition of a true slime mould, how the egg cell of a plant or animal initiates embryogenesis.

Life in the laboratory

Many of the points raised in the previous section would be only semantic, and most of the data presented in the following chapters would be unavailable, had it not been possible to culture *Physarum* in the laboratory from 'spore to spore'. For life-cycle studies, sporangia are crushed and a suspension of spores in water is spread on an agar surface. Amoebae will hatch, move around and engulf food provided as bacteria, e.g. *Escherichia coli*, either living or dead. They will divide and after a few days small plasmodia will be detected. They can be picked up after they have become a few millimetres in diameter and transferred to a new agar plate. They will grow very vigorously when fed autoclaved oat flakes, and can be subcultured about once every week or two. Where large masses of plasmodia are needed, they can be grown that way in a bath tub. To complete the life cycle, a growing plasmodium is put on an agar surface without food. After a few days on a window sill, fruiting bodies will form. These procedures have been described in detail by Gray (1938) and Collins (1979).

A breakthrough for biochemical studies on *Physarum* has been the establishment of a liquid medium that supports vegetative growth of plasmodia indefinitely, or so it seems (Daniel & Rusch, 1961). In contrast, plasmodia propagated on agar can grow old and may die after a few months. The essential ingredient in the axenic medium was, at first, chick embryo extract, but was later identified as the blood pigment haemin or haematine. In its semidefined state, the growth medium contains glucose, an amino-acid mixture (tryptone), yeast extract (providing vitamins), and a balanced salt solution, adjusted to pH 4.6. A fully defined medium is also available (Daniel & Rusch, 1962), but has not been much used as growth rate is one third of that observed in the semidefined medium.

Plasmodia are grown in two states: as a shaken suspension of microplasmodia of 1–3 mm in diameter, or as sedentary macroplasmodia where the upper surface is exposed to air. Typically, 20 ml stock cultures of microplasmodia are agitated in a 500 ml flask by a reciprocal shaker in the dark at 26 °C, and every 2–3 days 0.5–1 ml of the suspension is transferred into fresh medium, using sterile techniques. However, much larger volumes or containers and a fermenter have been used to produce large quantities of plasmodia and to allow the isolation of interesting molecules, such as ribosomal DNA and nucleic acid polymerases.

If microplasmodia are not regularly transferred after about one week, when the growth medium has become exhausted, most microplasmodia, or what is left of them, adhere to the glass surface forming a ring of 'spherules'. These can be suspended in sterile water, washed by gentle centrifugation, and placed onto a strip of filter paper using a pipette. When they have dried out they can be stored in the refrigerator for years, and after they are put

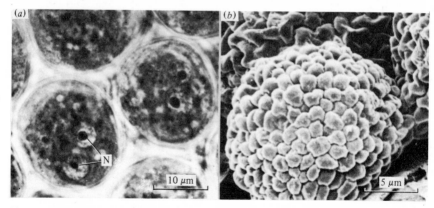

Fig. 7. Spherules, multinucleated walled cysts, a means of survival and a case of reversible differentiation. (*a*) Phase contrast micrograph, (*b*) scanning electron micrograph (R. Wolf, Würzburg). N, nucleus.

back on growth medium the spherules will germinate and resume growth. This method is very convenient for storing the various strains and spreading *Physarum* to laboratories all over the world. As a developmental pathway, spherulation may serve to survive adverse conditions in the field.

For the developmental biologist this process, which is also called macrocyst formation, is a case of reversible differentiation (Fig. 7). A better-controlled way to produce spherules is to transfer growing microplasmodia to a salt medium without nutrients, and spherulation will occur within two days rather synchronously on the shaker in the dark. Another means to obtain spherules is to culture microplasmodia in fully defined growth medium and then add manitol to a final concentration of 0.5 M.

Most biochemical work has been done on the macroplasmodia. They are produced by placing microplasmodia, which have been harvested with the centrifuge, on to a filter paper using a pipette. Within about one hour all microplasmodia have fused with one another (Guttes, E. & Guttes, S., 1964). (The filter was originally supported by a layer of glass beads in a Petri dish, but stainless steel grids are now in use.) Then growth medium is added, and the macroplasmodium will grow from 2 cm to about 10 cm in diameter within 24 hours. Such a macroplasmodium contains roughly 300 μg DNA, 3 mg of RNA, and 30 mg of protein. Cell biologists note that plasmodia fuse naturally, which allows for many experimental questions as discussed below. They have also discovered that plasmodia of the same species, but collected from different geographical locations, may not fuse. This raises the important quesions of recognition of self and incompatibility with strangers.

Provided that the microplasmodia are in exponential growth phase when setting up a macroplasmodium, naturally synchronous mitoses occur predictably every 8 h in the Petri dish kept at 26 °C in the dark, the minimum generation time.

The timing of mitosis can be checked by removing a small portion of the plasmodium and smearing it on a slide. After a brief fixation in a mixture of 50% ethanol and 50% glycerol, the mitotic stage can be readily deduced from the nuclear morphology in the phase contrast microscope (Guttes, Guttes & Rusch, 1961). Fig. 8 shows typical stages: (a) interphase, (b) prophase, (c) metaphase, (d) telophase and (e) daughter nuclei in which the nucleolus is being reconstructed from small particles. It is very convenient to know from the crescent-shaped nucleolus that the whole plasmodium is still 20 min away from metaphase which is a very accurate reference point, as it lasts only 5 min. As frequent sampling of pieces from a macroplasmodium does not seem to interfere with nuclear division, it has been ascertained by looking at smear preparations from all over a single macroplasmodium, and at all nuclei isolated from each sample, that metaphase occurs in almost all nuclei (more than 98%) within 5 min after the second and third post-fusion mitosis in plasmodia up to 7 cm in diameter. This enables experimenters to work with multiple samples of homogeneous material from a single 'cell', and by cutting a plasmodium into pieces like a cake, each sample can be handled with ease by holding it at the edge of the supporting filter and transferring it to other media, containing drugs or radioactive precursors, or to other plasmodia for fusion experiments.

Just as microplasmodia encyst under adverse conditions, a macroplasmodium will transform into a crust, a sclerotium. It breaks up into many pieces. They become walled macrocysts, probably identical to spherules, which contain several nuclei (up to 15). They are held together by a sheath which is made up of masses of a fibrillar network of polysaccharide material, the dried 'slime'. As many cysts are formed in the absence of exogenous nutrients, a drastic change must occur in the metabolic pathways to enable the production of much carbohydrate from endogenous sources.

As a macroplasmodium is a synchronous organism, it would be a better starting material for the analysis of macrocyst formation than microplasmodia, which are synchronous with respect to nuclear division cycles of several hundred nuclei within each individual, but out of phase with each other. However, a macroplasmodium in the process of sclerotisation constantly changes its shape during its migratory phase and cannot be adequately sampled over the many days which are required for this differentiation process.

Therefore, the recent observation, that one single macroplasmodium can directly be transformed into many macrocysts within one day in a completely synchronous fashion on salt medium plus 0.5 M mannitol, is an important

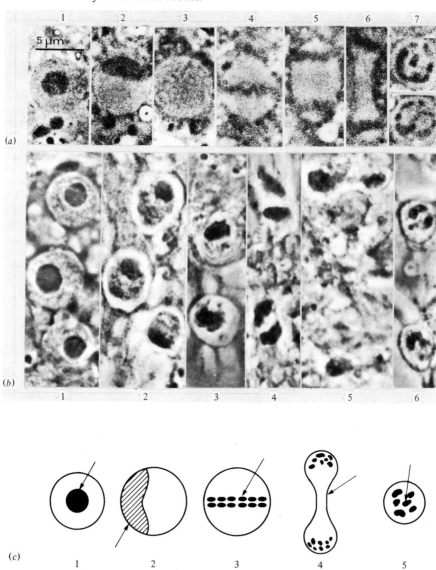

Fig. 8. Mitosis in *Physarum*. Some stages of nuclear division as seen: (*a*) *in vivo* (from a film analysis by R. Wolf, Würzburg); (*b*) after fixation (phase contrast micrographs by R. Wolf, Würzburg). (*c*) Some prominent markers (arrows), useful for staging mitotic events for cell cycle studies: 1, interphase (one central nucleolus); 2, crescent-shaped nucleolus, 15 min before metaphase; 3, metaphase; 4, dumb-bell stage (end of telophase, 2 min after metaphase); 5, nucleolar reconstruction (10–45 min after metaphase).

improvement for the analysis of this developmental pathway (Jalouzot & Toublan, 1981).

Sporulation, the alternative pathway which is essential for the life cycle of *Physarum*, cannot be analysed unless it can be accurately predicted. A marker, such as the crescent-shaped nucleolus for mitosis, is not available, and when sporangia are observed one fine morning, it is too late to ask what really happened. Fruiting bodies are usually produced at night and not during the day. This fact alone indicates that specific conditions may be required for this irreversible differentiation process.

One necessary requirement for sporulation in *Physarum* is illumination, as first shown experimentally by Gray (1938). On the other hand a well-fed plasmodium will only sporulate if it has been starved for several days. A very reliable procedure to obtain sporulation predictably and synchronously, not only in one but in dozens of macroplasmodia with up to 100% success, has been devised by Daniel (1966). Microplasmodia must be harvested at the end of exponential growth phase and allowed to fuse on a filter paper. The ensuing macroplasmodium is then kept for exactly 4 days in the dark on non-nutrient sporulation medium containing niacin as an indispensible component. Then the plasmodium is illuminated for 4 h. By about 8 h after illumination the plasmodial strands break up into beads about 1 mm in diameter. Each bead becomes segregated into a stalk, with most of the pigment, and a head. The protoplasm within the head is divided up into mostly uninucleated portions, the pre-spores (cellularisation). Finally, by 18 h after illumination the sporangium has turned black due to the production of a dark pigment, melanin, and the sporulation process is complete (Fig. 9). Unfortunately, most strains of *Physarum* do not respond with such precision if handled in this manner.

An intriguing question for the developmental biologist concerns the role of light. As spherules and spores are both unicellular dormant forms for survival and propagation of the organism, there may be one common developmental pathway leading from growth through starvation to spherulation (in the dark) and sporulation (in the light). This would make both end-products quite similar. On the other hand, a starved plasmodium may have only two options: either to grow (if fed) or to differentiate. If it is not fed, it has again two options: either to make spherules or to make fruiting bodies. The light could activate a switch, in that spherulation is now blocked and the sporulation programme is selectively activated. This alternative pathway leads to commitment, photomorphogenesis, cytodifferentiation into spores and, as we shall see below, meiosis. Consequently, sporulation could provide us with a model to analyse the major decisions in the life cycle of any higher organism, i.e. determination and differentiation (Sauer, 1973, 1974, 1980).

Despite the problems encountered in setting up the complete sporulation

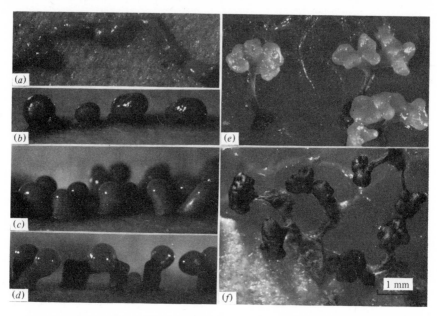

Fig. 9. Sporulation, a case of morphogenesis (photos by R. Wolf, Würzburg).

regimen, the developmental reactions which follow commitment to sporulate can be reliably analysed if well-fed plasmodia on filters are transferred to a growth medium of one-tenth normal strength (or on 2–4% agar, made up with the dilute growth medium or sporulation medium), kept for various periods of time and illuminated repeatedly for various lengths of time. After any of these treatments plasmodia will eventually segregate into beads and form sporangia.

In each case of sporulation, one light-induced presporangial mitosis has been observed, from the time that *Physarum* was grown on microorganisms (for review see von Stosch, 1965) to the most elaborate cytological analyses (Guttes, Guttes & Rusch, 1961; Laane & Haugli, 1976). It is a curious fact that a starved plasmodium which has consumed all its reserves and is close to dying, undergoes that mitotic nuclear division, and some developmental biologists speculate that along with the commitment to differentiate into something really different a specific mitosis is required for any stable differentiation process to take place.

On the other hand, despite this nuclear division, and previous ones at 24–36 h intervals during starvation in the dark, the number of nuclei per plasmodium precipitously declines by a process of nuclear degradation and elimination which is a characteristic of the life of myxomycetes.

The occurrence of meiosis has been generally accepted although its exact

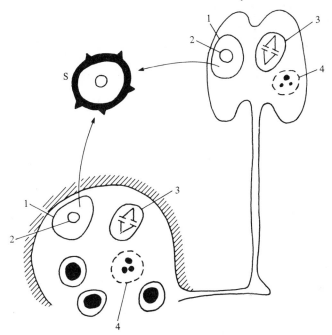

Fig. 10. Timing of maturation divisions (meiosis). Cellularisation (1) and meiosis (2), as well as presporangial mitosis (3) and nuclear elimination (4) take place in the sporangium (*right*), but can also occur in the plasmodium (*left*). As a result spores (s) are formed.

time and manner have been hotly debated for decades. Much of the controversy has been resolved, and even a new insight into developmental pathways is gained if we accept two ideas: (1) that the timing of meiotic events is not rigidly controlled and (2) that meiosis is optional. This conclusion comes mainly from the comparative cytological studies and the mating behaviour described in the following chapter.

Meiotic events can be inferred from the detection of synaptonemal complexes in the electron-microscope (indicative of meiotic prophase), by comparing chromosome numbers and the DNA content (biochemically or in the cytophotometer), and by the segregation of genetic markers in the progeny. None of these parameters is particularly easy to study in *Physarum* but all are essential to get a clear picture.

When all the evidence is weighed (Fig. 10), we learn that in most cases meiosis occurs after the sporangia are formed, and that two divisions take place in the spore within about 24 hours. Also, after each division one nucleus is aborted. This leads typically to uninucleated mature spores. However, meiosis can also occur in the plasmodium, after the presporangial mitosis but before morphogenesis has begun. Also, two to four nuclei can remain in the spore when, rarely, the products of meiotic divisions survive, and finally, more than two nuclei may be in the pre-spore, and both undergo meiosis together (von Stosch *et al.*, 1964; Laane, Haugli & Mellum, 1976).

In some strains of *Physarum*, even though synaptonemal structures are visible and nuclear divisions do occur in the fruiting body, the same number of chromosomes has been counted in the nuclei of the plasmodium and the spore. It seems that some sort of pseudo-meiosis is necessary to go from the multinucleated plasmodium to the unicellular stage, and this holds for life cycles that may be continually haploid or continually diploid.

Although the atypical events surrounding meiosis are quite rare, they can have far-reaching consequences for our understanding of the life cycle, if neither morphogenesis nor meiosis is absolutely required.

A brief look at a distant relative of *Physarum*, although still a bona fide acellular slime mould, may be instructive. In *Ceratiomyxa*, after the pre-sporangial mitosis at the periphery of the plasmodium, many single spores are formed, each sitting on a tiny stalk. They are not enveloped in a fruiting body (which classifies *Ceratiomyxa* as an exospore myxomycete, and *Physarum* as an endospore myxomycete), and each contains one nucleus which gives rise to four nuclei after two normal meiotic divisions. A tetranucleated cell germinates from the spore, and its nuclei divide immediately in synchrony to give rise to eight swarm cells (see von Stosch, 1965).

The microscopic progeny of a fruiting of *Physarum* can be grown on a solid substrate, just like macroscopic plasmodia. In both instances this is a form of vegetative proliferation. However, the requirements are quite different: no axenic medium is available to grow these cells efficiently on a solid substratum like agar, although food bacteria can be killed with formalin. Consequently, the amoebae contain foreign macromolecules in their food vacuoles which complicates biochemical comparison of the two life-forms of *Physarum*. This problem has been overcome, as one complex (Goodman, 1972) and one less complex medium (McCullough & Dee, 1976) have been devised in which amoebae will multiply axenically, and a detailed comparison is now possible.

When a very dilute suspension of spores and bacteria is spread over the agar surface, clones of amoebae can be obtained. One regular observation is a mitotic division which occurs just a few hours after germination. Actually, two kinds of cells can hatch from a spore, either a non-polar amoeba or a swarm cell. The latter has two flagella, a long one and a short one, which are inserted in a cone containing two basal bodies. The cell nucleus is firmly attached to this cone, giving the swarm cell a more rigid polarised shape than the amoeba. Amoebae can be quantitatively transformed into swarm cells just by suspending them in water. This process takes from 30 min to 3 h in various strains of *Physarum* and can be viewed as yet another differentiation process, displaying not only a completely different motility but also considerable subcellular morphogenesis. This process is readily reversible, and swarm cells will revert to amoebae on an agar substrate lacking free liquid.

Amoebae move around, eat bacteria, and divide by fission of their bodies after typical 'open' mitosis. Flagellated forms have to retract their flagella and use the polar basal bodies as centrioles for their mitotic spindles. As amoebae multiply, an area free of bacteria arises in the lawn of food bacteria, a plaque which can be easily spotted and serves to identify a clone.

Closer inspection of plaques indicates that amoebae display two kinds of behaviour: a spreading movement and an aggregation. The former leads to an expansion of the cell population as a monolayer, the latter to a typical ring structure made up of thousands of amoebae bordering on the bacterial lawn. Within the ring some of the amoebae have rounded up and become cysts. These are called microcysts to distinguish them from the multinucleated 'macrocysts' that are derived from plasmodia (Jacobson & Dove, 1975). A whole suspension of swarm cells or amoebae can be forced to encyst by unfavourable conditions, and this process is readily reversible by adding growth medium. This enables developmental biologists to analyse yet another differentiation process of *Physarum* in a homogeneous cell population.

Microbial genetics can be very effectively applied to a clone growing from a single amoeba. Large numbers of genetically identical individuals can be handled and mutagenised. Some interesting mutants have been discovered by extensive screening or employing clever selection techniques (Dee, 1973), some of which enable us to dissect pathways of development, as we shall see below.

As the amoebal stage of myxomycetes is haploid, mutants occurring in the natural habitat will be immediately expressed. Thus, a rare favourable mutation can be selected by a change in the environment. Progeny from such an individual could either immediately outgrow the other members in a population or have a selective advantage after a long, dormant state as a microcyst. Whether this mechanism works in nature to adapt slime moulds to very different or catastrophic changes in the environment, or creates new species at a high rate, must be established by field studies. This may not work as efficiently as it seems, as amoebae do not normally grow in clones but form plasmodia which no longer consider amoebae as brothers or sisters, but eat them alive, even if encysted, leaving no chance for the 'fittest' of them to survive. In the laboratory it is by such drastic changes that drug-resistant or heat-sensitive (conditional) mutants are selected.

After these observations we can expand the life cycle of *Physarum* to include the optional pathways of differentiation which can operate in the microscopic unicellular or in the macroscopic multinucleated stage (Fig. 11). We can distinguish two alternate states of vegetative proliferation: amoebae

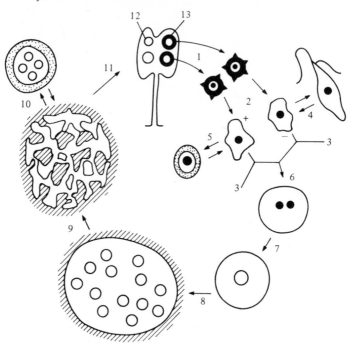

Fig. 11. The full life-cycle of *Physarum*. From spores (1), amoebae of different mating type (+ or −) germinate (2) and either proliferate (3) or differentiate into a flagellate (4) or microcyst (5), or they fuse (6) to form a zygote (7). The diploid zygote transforms into the plasmodium (8) (amoeba–plasmodium transition) which can grow or, if starved (9), differentiate into macrocysts (10) or sporangia (11); pre-spores form in the head of the sporangium (12) and, after meiosis, become haploid spores (13).

or plasmodia; one bipolar motile form: the swarm cell; three dormant forms: microcysts, macrocysts (spherules), and spores; three examples of reversible differentiation: microcyst formation, spherulation, and amoebal–flagellate transition; two pathways which display irreversible commitment: sporulation and amoeba–plasmodium transition. Thus, the life cycle of *Physarum* reveals a truly multipotent system: the big, yellow, slimy plasmodium with its shuttle streaming, closed mitosis (i.e. nuclear membrane persists) and lack of cellularisation, has the potential to grow, or to turn into either macrocysts (spherules) or fruiting bodies (sporangia). The unicellular stage can develop in one of four ways: grow as amoebae or swarm cells, stop growth and encyst, or become a gamete. The latter does not look any different but is a remarkable cell as we shall see in the following section.

The mating game

As we already know that meiosis occurs during sporulation, the unicellular stage must be haploid. Consequently, amoebae as gametes must fuse before a plasmodium can arise. Cell fusion (i.e. syngamy) is not enough to establish the diploid state; nuclear fusion (i.e. karyogamy) must follow. Such is typical in all sexual life cycles where gametes fuse before a diploid phase of development sets in, and after the zygote has formed. Fungi are somewhat different in that nuclear fusion can be delayed for long periods, as they are dikaryotic. The proportions of the haploid and the diploid state of an individual life cycle vary considerably, but it is clear that in higher organisms most of life is spent in the diplophase, and the haploid gametes, eggs and sperm, do not multiply once formed. They are also short-lived and will die if they do not meet quickly. When they fuse, one sperm with one egg, this process of insemination serves two functions: to activate development, and to recombine two different genomes, one from the father and one from the mother. In some cases the two functions can be separated. As an example, the mature frog egg can be activated by pricking with a needle, and a frog may develop which contains only the maternal genome. In many animals and plants this form of parthenogenetic development occurs naturally, and asexual life cycles can regularly alternate with sexual ones. Some have debated whether there is any use for sex, in the evolutionary sense of meiosis and gene recombination (Maynard Smith, 1971; Dawkins, 1976).

In any event, when a multicellular organism develops from a single cell (either following fertilisation or parthenogenetically) it begins differentiation very soon: plants decide on root and shoot, animals on the germ line and body cells (the soma). In every case the complex organism derives from a large and complex cell, the egg. These cells have undergone a long and extensive development called oogenesis, and some essential steps of embryonic development are pre-programmed. Such is evident alone from the terms of polarity, localised cytoplasmic determinants, maternal mRNA of enormous complexity, and informosomes (Davidson, 1976). Most of the pertinent – and unsolved – problems in developmental biology of multicellular organisms stem from the mysterious phase of oogenesis. A sperm cell is terminally differentiated; it is highly specialised to function for fertilisation, yet it cannot develop into an embryo.

One prerogative for the evolution of higher organisms from single cells is the transition to multicellularity. It is clear that the division of labour, later to be observed in different tissues and organs, may already be preconditioned by the invention of the egg cell. Therefore, the formation of a plasmodium in true slime moulds, from its size alone, is more than a unicellular organism, but undifferentiated except for one single fact: it is definitely no longer the unicellular organism it was before. Consequently, the development of a plasmodium offers a unique possibility to analyse one step in the establishment of multicellularity whereas cellularisation and further differentiation only occurs later on in its life cycle.

How does a plasmodium arise? Once again, as in the timing of meiosis, we encounter a great flexibility. Fusion and zygote formation cannot be easily observed, as it cannot be predicted when an amoeba or flagellate will decide to become a gamete, and only a few ever do so in a population. But occasionally these events have been seen or filmed. Plasmodia arise in the middle of a plaque of amoebae, so maybe a gamete no longer multiplies or moves as fast as the other amoebae that make up the ring-shaped aggregate of a population. If you start with a single amoeba and look at the clonal population, it is quite normal that no plasmodia are formed. It seems that genetically identical individuals cannot fuse and start a diploid plasmodium.

If you start many amoebal clones from one fruiting body, and mix a few amoebae from two separate clones, you will notice that plasmodia arise on average in every second mixture. If you pick up several plasmodia, let each grow until it finally sporulates, and do the same kind of crossing with the progeny, you will again get a successful 'mating' in 50% of the crosses. As identical amoebae do not mate, those that do must differ in at least one characteristic. Since the result of random mixing is always 50% mating, it can only be one single factor that is different. The same result is observed if amoebae from the previous generation and the following generation are mixed: consequently, it is the same factor that is required for mating in consecutive generations. Obviously, there is genetic control over mating, and these developmental events are most easily explained by assuming that one gene, the mating-type gene, exists in two states (true alleles), and that both of the functions are required in the plasmodium, while either one satisfies the needs of the amoeba. (This argument can be turned around: it could be that the mating-type function represses plasmodium formation in the amoeba, and that the two allelic, though different, functions cancel each other out when they are expressed in the same individual cell, in this case the zygote.) Generally speaking, even though gametes look alike, they are not (this situation is called heterothallism) (Fig. 13a). At first sight they behave like male and female: one has the mating type mt_1, the other has mt_2. Then the diploid plasmodium will have both mating-type genes (mt_1/mt_2), and following meiosis the haploid spores will be either mt_1 or mt_2 in equal proportions. Consequently, the fertility of crosses is 50%, and a cross is required to complete a life cycle (Fig. 12).

This is only part of the story, as becomes obvious when a specimen of the same species of *Physarum*, yet from a different geographical location, is analysed. Just as in the first case its cross fertility (i.e. crosses with amoebae derived from one fruiting body) is 50%. However, when amoebae of *Physarum* from the two locations are crossed with each other, cross fertility

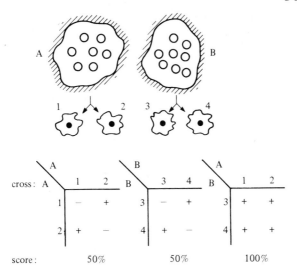

Fig. 12. The multiple mating types of *Physarum*. Plasmodia from two different locations (the isolates A and B) produce amoebae of two mating types, designated 1 and 2 for isolate A, or 3 and 4 for isolate B. The crossing of amoebae from isolate A or B yields 50% cross-fertility (score), but crosses betwen A and B are 100% cross-fertile. Consequently, the mating types 1, 2, 3, 4 are different alleles of one mating gene.

is 100% (Fig. 12). This result is surprising, as one individual seems to be a 'male' in one location and a 'female' in the other. It can be explained by assuming that in different locations we will find a population of *Physarum* that has two alleles of one mating-type gene, but the pair of alleles is different in each population. In this case amoebae of mt_1 or mt_2 and mt_3 or mt_4 will always yield fertile progeny as no two identical alleles will be combined in successful matings (Fig. 12). So far, in each of 14 different locations where *Physarum* has been collected, a different pair of alleles of the mating type has been distinguished.

In genetic terms, this situation is called a multi-allelic single mating-type locus, and the great variety in alleles of one gene indicates perhaps a high rate of 'mutability'. One wonders, however, why only two alleles are expressed in one population. At any rate, heterosexual slime moulds are not just 'boys' and 'girls'.

After fusion of two amoebae of the appropriate mating type, and their nuclei, it is hard to decide when they have become committed to be a plasmodium. Once the nucleus divides and cell division is left out, we can be sure that a plasmodium will form by consecutive synchronous nuclear divisions. This would give rise to a pure plasmodium (by repeated nuclear divisions within a single cell). But this is not the way it works. Several plasmodia, which arise in a crossing, can afterwards fuse with each other. Such fusion is typical for syncytium formation. Consequently, the multi-nucleated large mass of protoplasm of *Physarum* is neither a true plasmodium nor a true syncytium, it is a unique form, but let us settle for 'plasmodium', as it is an essential stage and the first step in the developmental sequence.

After this description, we must now distinguish two kinds of cell fusion: a sexual fusion for mating, and a vegetative fusion. While sexual fusion requires that the partners be different, all young plasmodia can fuse with one another, i.e. identical cells fuse. As we shall see below, vegetative fusion is also under genetical control, and only plasmodia with identical fusion-control genes do so.

This causes a problem for the analysis of sexual mating. It is known that young plasmodia can fuse with big ones, which can result in a mixture of different nuclei in one plasmodium (a heterokaryon). On the other hand, there must be a fusion barrier for identical amoebae, otherwise they would not exist (the experimenter can distinguish plasmodium fusion from gamete fusion by choosing appropriate fusion genes). This is not a trivial observation, if we look at it together with the next form of beginning of a life cycle in *Physarum*.

It is quite frequently observed that single amoebae, derived from certain strains of *Physarum* (or a relative such as *Didymium*) can be cloned, i.e. they yield plasmodia regularly. In this case identical cells have supposedly mated, and this has been called homothallism (Fig. 13*b*). Consequently, the life cycle would consist of haploid and diploid phases, both with completely identical genomes. In many cases amoebae from such strains will not cross with amoebae from any other strain. This indicates that, although it is a single species, *Physarum polycephalum* can occur in nature as various groups that will not interbreed (i.e. which does not follow the definition of a species any more). It is not clear whether the putative homothallic or the bisexual life cycles prevail in the myxomycetes in their natural habitat. One problem with homothallism in *Physarum* is that it may not exist at all. In all cases where plasmodia can be derived from one amoeba, i.e. clonally, it has been demonstrated that the chromosome number is the same throughout the life cycle. Consequently, no fusion of two amoebae and their nuclei has occurred to initiate a plasmodium (Fig. 13*c*). Therefore, strains with this ability have been termed 'selfer' (or apogamic, or asexual). This is an important result, as it proves that neither sexual fusion nor nuclear fusion is required for plasmodium formation, and consequently, the transition from a plasmodium to spores does not need a complete meiosis. This mechanism of selfing operates in all forms of myxomycetes, and in one creature, *Echinostelium*, which produces only minute plasmodia, sexual fusion has never yet been observed. In this species, the tendency of somatic fusion is

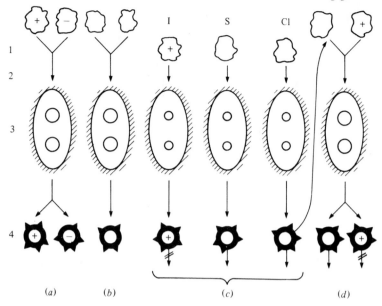

Fig. 13. Various modes of plasmodium formation. There are two kinds of plasmodium formation, either with (*a,b*) or without (*c*) fusion of amoebae (and change in chromosome number, ploidy). (*a*) If the amoebae that fuse are of a different mating type (either + or −), sexual fusion (1) occurs before amoeba–plasmodium transition (2), a heterothallic plasmodium (3) is formed and its progeny (4) consist of two kinds of spore (+ and −); this defines heterothallism. (*b*) If genetically identical amoebae fuse, a homothallic plasmodium is formed, and all progeny are identical. (*c*) If an amoeba can turn directly into a plasmodium, apogamic development has occurred, not requiring sexual fusion. Three kinds of such plasmodia have been distinguished: the illegitimate plasmodium (I) whose progeny cannot self (broken arrow), the selfer (S) whose progeny amoebae can self (arrow), and strain Colonia (Cl) whose progeny has the option to either self (arrow) and remain haploid, or fuse with a heterothallic amoeba and become diploid, with progeny half of which can self, half cannot (*d*).

also repressed while the small plasmodium tends to divide in two, regularly. This form of vegetative propagation is called plasmotomy, to distinguish it from cell division proper. Such a non-sexual life cycle has been defined as apogamic (Fig. 13*c*).

An example of 'selfing' occurs naturally in one *Physarum* strain, called 'Colonia', which has become the pet organism for much subsequent genetical analysis (Dee, 1973). Here, a clonal population regularly produces plasmodia. It has also been discovered that amoebae of Colonia can be crossed with practically any other *Physarum* strain (Fig. 13*d*). Obviously, Colonia possesses a special mating type that allows both selfing and crossing. At first, it was suspected that selfing occurred by fusion (i.e. Colonia is homothallic). Later it was found that the chromosome number does not change, and even though synaptonemal complexes are observed in pre-spores, a pseudo-meiotic division yields viable haploid spores, although not very many (Laane *et al.*, 1976). Consequently, the Colonia strain is apogamic, and a single amoeba can turn into a plasmodium, a process that has been clearly demonstrated in a recent film (Holt, 1980). As the same strain can also be heterothallic if allowed to cross, and produces equal numbers of progeny with either the special mating type (Cl) or whatever mating-type allele went into the crossing, the optional differentiation of an amoeba into a plasmodium with or without sexual fusion is dependent on the 'mating-type gene'. The analysis of developmental mutations given below stresses the involvement of this polymopth single-gene locus which is really a developmental control gene, as it does not control sexual fusion.

This raises the question of the stability of pure clones of heterothallic amoebae and brings us to the last type of plasmodium formation, which occurs in a clone of any amoeba, if one only waits long enough. This kind of plasmodium formation has been classified as 'illegitimate' because it is not supposed to happen (Fig. 13c). As the progeny of such illegitimate parents have not changed their original mating type, their children are again heterothallic (i.e. to become a plasmodium they must mate – rapidly) one would have to assume that the supposed mutational event that has initiated the plasmodium in the first place does not persist. An easy way around this problem is that such mutation is not stable in the plasmodium, and only 'normal' spores give rise to viable offspring. This sounds rather speculative, but nuclear elimination occurs in all myxomycetes, and as we shall see, the clear-cut genetic analysis, obtained with strain Colonia and the murky heterogeneity of chromosome numbers in nuclei of 'isogenic' plasmodia, can only be explained by a stringent nuclear selection process operating at sporulation.

As summarised in Fig. 13, we have distinguished several possible ways of amoeba–plasmodium transition. (*a*) In most instances the strains are heterothallic, gametes require sexual fusion, and two alleles of a multi-allelic complex genetic locus are involved in establishing the plasmodium. (*b*) If homothallic forms exist, they arise after sexual fusion of completely identical partners. (*c*) In three cases plasmodia will form in clones derived from a single amoeba: either illegitimately and very rarely (I), or selfing occurs regularly (S). In this situation there is no change in ploidy, and the whole life cycle is haploid (or diploid). Among these selfers is strain Colonia (Cl) which displays an alternate mechanism of either sexual or asexual life cycle which may facilitate the unravelling of an important developmental switch.

What is so exciting about the Colonia strain? Its life is haploid throughout the cycle from spore to spore. Hence most mutations produced in the ameobal state show up immediately, even in the plasmodium. Those that do not, might lead to the detection of essential genes that regulate the initiation and maintenance of the plasmodial state. Progeny from haploid plasmodia can be cloned and mutagenised in large numbers – as mentioned above – but as they will cross with other amoebae in sexual mating, pure Mendelian genetics are possible, and any gene can be crossed into the Colonia genome to construct strains with an array of selected markers. Linkage groups can be established and gene distances calculated from recombination frequencies. It can be assumed that a combination of biochemical and genetical analyses will provide hard facts about the life of *Physarum*.

In summary, the mating game of *Physarum* is quite variable, and in some instances progeny are obtained without mating at all (in the selfers).

Let us now attempt to sketch how a myxomycete life cycle may have evolved, by taking up some views of biosystematics (Collins, 1979). All myxomycetes collected in nature can occur as either sexual (heterothallic) or selfers (non-sexual or apogamic forms), and even though no sexual fusion occurs in *Echinostelium*, it is not decided which of the two modes of plasmodium formation arose first in evolution, as *Echinostelium* could be either very old, or it has only recently reverted from the sexual to the asexual state.

At any rate, we can start from a unicellular state as shown in Fig. 14, the amoeba-flagellate. As long as these cells divide or encyst, their life cycle is quite simple. Next, an uncoupling of nuclear division and cell division leads to the invention of plasmodial organisation and a large free-living organism. The plasmodium can grow, break up by plasmotomy, as another means of vegetative propagation, or encyst as a macrocyst and survive unfavourable conditions (*Echinostelium*). If one starts with a large plasmodium, it could give up its acellular state and make cells without much precision in counting nuclei; the macrocyst might originate this way.

With the introduction of meiosis at the time of cellularisation of a plasmodium and the establishment of the ensuring meiocyst, which yields one or more uninucleated progeny, the plasmodial and unicellular life cycles become joined together. While the macrocyst can directly return to the plasmodial state, the meiocyst can only become a plasmodium by amoeba–plasmodium transition since it leads to uninucleated cells. If meiosis is complete, mating of haploid heterosexual amoebae will yield a zygote; if not, apogamic development will lead to a plasmodium more directly (without ploidy change). In any event, with the arrival of the meiocyst, the life cycle contains an irreversible commitment, as the hypothetical meiocyst cannot directly turn into a plasmodium. The fruiting process of *Ceratiomyxa* mentioned above, may illustrate a case where meiosis is initiated in the absence of morphogenesis.

The third line of development, besides integration of a plasmodial organisation with a meiocyst, is characterised by morphogenesis. Thereby a small plasmodium (like that of *Echinostelium*) is transformed into a single or a

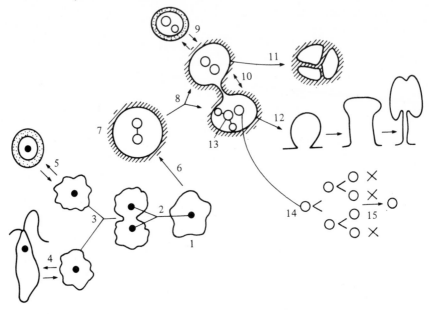

Fig. 14. A sketch of possible evolutionary pathways open to myxamoebae. An amoeba (1) proliferates by open mitosis (2) followed by cell division (3) as a micro-organism, and can differentiate reversibly into a flagellate (4) or a microcyst (5) – cytodifferentiation. Alternatively, a macroscopic creature, the plasmodium, arises by amoeba–plasmodium transition (6), involving closed mitosis and uncoupling of nuclear from cell divisions (7). The plasmodium can grow and divide by plasmotomy (8) or form cysts (9). The two plasmodia can either fuse again (10) or cellularise (11), or undergo morphogenesis, leading to fruiting bodies (12). The nuclei undergo either proliferative (13) or meiotic divisions (14), leading (in most cases) to a haploid genome (15). In *Physarum*, under normal conditions, the various pathways and events are coupled and integrated in the life cycle shown in Fig. 11.

large plasmodium (like that of *Physarum*) with many fruiting bodies, involv-ing extensive cellular movements and shaping. It seems that this process is another irreversible differentiation step, and by its temporal coupling with the development of meiocysts (now called prespores) and their differen-tiation into mature spores in the elevated head of the sporangium, the full life cycle has been stabilised, now displaying two points of commitment: sporulation and amoeba–plasmodium transition.

The flexibilities in timing of developmental events observed in the life cycle of *Physarum* in the laboratory, and all other myxomycetes outdoors, lend some credence to the evolutionary speculations outlined above. They illustrate that separate lines of development – mitotic proliferation, meiosis, cytodifferentiation, morphogenesis, and transition beyond the microscopic unicellular form of life – are coupled and integrated. However, they can be dissected experimentally in *Physarum* as in no other higher organism. We shall limit our discussion to selected experiments in developmental, cellular and molecular biology, and attempt to show that *Physarum* is a model system, even though today complete answers cannot be given.

A puzzling phenomenon is the variability of chromosomal numbers in any plasmodium, in those freshly cloned from a single amoeba even more than in those kept for several years: chromosomes vary from 20 to 200, the haploid set being about 40. Although the clean genetic results are explained if an elimination of all atypical nuclei has occurred in the viable spores, biochemical work could lead to less conclusive data than expected. However, it could also be argued that the nuclei are partially polyploid, and the genetic make-up of a plasmodium may be more homogeneous than it looks from chromosome numbers. Even if they have still to improve *Physarum* as an experimental system (as a living thing it seems to have done quite well for millions of years) the 'Physarologists' tend to look down upon tissue culture cells in their randomness and abnormal chromosomal make-up, and stick to their favourite mould.

Nevertheless, some aspects of development, as projected from the lives of *Physarum*, may be instructive beyond level of moulds. Indeed, the main incentive to work with *Physarum* as a model was to find a cure for human cancer (Rusch, 1959, 1969, 1970).

2

Of man and moulds

Many biologists have come in contact with slime moulds. Some are curious about the life of this group of organisms, others have had specific biological problems in mind, which they have hoped to tackle experimentally, employing moulds as a model. Somehow these contacts have been contagious, and *Physarum* has spread from a few centres to laboratories all over the world.

To provide rapid communication, the *Physarum Newsletter* – with a yellow title page reminiscent of the still unidentified yellow pigment of *Physarum* – appears about twice a year. It is edited by T. E. Evans and contains useful information, references to work in press and print, and a mailing list of friends of *Physarum* that is rapidly approaching 200. A recent issue contains close to 100 new references, and the topics cover a very wide range of interests. This microcosm about *Physarum* – and a few related genera of true slime moulds like *Didymium* and *Echinostelium* – comes to life each year at the *Physarum* conferences which are held alternately in the USA and Europe at various locations. On two occasions a cross-section of current work has been published: a collection of reviews by Hüttermann (1973), and numerous individual papers and abstracts by Sachsenmaier (1979). While the mailing list provides an almost full account of the current 'Physarologists' and the newsletter shows the diversity of projects, an attempt will be made to trace coherent lines of research by listing some people and their projects, as well as the strains of *Physarum* that are kept in their laboratories. These are examples of true integration within the international scientific community.

The person who named slime moulds myxomycetes was Wallroth (1833). This was some time after Schweinitz (1822) had called *Physarum* by name, and a long while before this creature was classified as a true slime mould by Jahn (1928), with alternating sexual and vegetative phases. The biology of myxomycetes has been carefully described by Alexopoulos (1960), Gray & Alexopoulos (1968), and Martin & Alexopoulos (1969). Their taxonomy has been treated by Eliasson (1977), and an authoritative book on the acellular slime moulds has been written by Olive (1975) who called them 'Mycetozoans'. A very readable short introduction to *Dictyostelium*, the cellular slime mould, and *Physarum*, the acellular slime mould, is available (Ashworth & Dee, 1975), and an outline of the potential genetic approach to

Physarum is provided by Dee (1975). The general biology of myxomycetes in the early days is summarised by Martin (1940), and an excellent review on the developmental physiology has been compiled by von Stosch (1965).

Originally, the true slime moulds were not recognised as unique model systems. Biology students were allowed a brief look at a plasmodium, if at all, as an example of pure protoplasm creeping around like a giant amoeba and displaying intensive protoplasmic streaming. Besides, there was a better known, albeit cellular, slime mould, *Dictyostelium*, and the feeling prevailed that if you have seen one you have seen them all. However, people in three areas of biology have become interested by the lives of the myxomycetes.

First, there has been an undisclosed relationship among amoeba, flagellate, plasmodium, and fruiting bodies, and von Stosch may be credited with the understanding of the life cycle where two sexually distinct gametes (+ and −) fuse and yield a diploid plasmodium which, by the fruiting process and meiosis, leads to millions of haploid spores. He was also the first to detect a great variety in the meiotic divisions and even an asexual life cycle (apogamy) which was hotly debated for many years. Other people involved in the analysis of the microscopic stages are Ross (1966), Aldrich (1968), and M. R. Henney & H. R. Henney (1968). The complex life style has been admirably rationalised in a recent review on the biosystematics, as evidence for evolutionary mechanisms creating reproductive systems (by Collins, 1979).

The next area covers a section of cell biology, motility, which has been investigated ever since Vouk (1910) observed the protoplasmic shuttle-streaming. Earlier observations, including the history of motility studies in *Physarum*, have been reviewed (Komnick, Stockem & Wohlfarth-Bottermann, 1973). Ingenious experiments were initiated by Kamiya (1959) to measure the motive force on isolated plasmodial strands in a pressurised double chamber. Wohlfarth-Bottermann (1979) has orchestrated a comprehensive effort on the ultrastructure, physiology, and search for the endogenous oscillator, and given a definitive summary of the work in his laboratory. Much of the current work on molecules and models, including that of Nachmias & Kessler, is contained in two recent books, edited by Pepe, Sanger & Nachmias, 1979, and Hatano, Ishikawa & Sato, 1979. The fact that the amino-acid sequence of actin from *Physarum* is almost identical to that of *Dictyostelium*, but also very similar to the cytoplasmic actin in mammalian cells (Vandekerckhove & Weber, 1978), may have two relevant implications: first, as this molecular structure is highly conserved in evolution, results obtained with *Physarum* as a model system may be of general significance; and second, the cellular slime moulds may not be all that far removed from the acellular ones in the evolutionary tree.

The directional movement of a 'hungry' plasmodium, while faster than that of regular amoebae or myxamoebae, can be best described as 'amoeboid' (Lewis, 1942). This means that endoplasm is pushed from the rear to the front, where it is in part transformed to the stiff ectoplasm. This process, superimposed on the shuttle streaming, together with slime secretion, leads to efficient migration over long distances, only to be interrupted for the brief period of mitosis (Anderson, 1964).

The question of why a plasmodium should be going anywhere, has been asked by Carlile (1970) in his studies on chemotaxis. At first, it looked as if *Physarum* could select those substances that it likes to feed on, sensing a concentration gradient, but later work has shown that some attractants cannot be metabolised, while some food-stuffs, if too much is offered, become repellents (Chet, Naveh & Henis, 1977). Since then, more direct methods for the chemotactic response have been applied, and changes in the motive force (Ueda & Kobatake, 1978), in surface charge, and a conformational change of the outer membrane were detected as a chemotactic substance reaches the threshold value (Hato *et al.*, 1976).

One general point can be raised: besides specific receptor molecules, such as perhaps those for glucose, an attractant (or repellent) may alter the orientation of water molecules at the cell surface, and this alteration is somehow transmitted to the contractile apparatus. A similar mechanism may act in our white blood cells; consequently, an understanding of the motile behaviour of *Physarum* may shed some new light upon such unrelated mechanisms as defense reactions of the blood system and smooth muscle contraction.

From another angle, mutants of motility, plaque morphology and amoeba–flagellate transition have been detected (Kerr, 1960; Jacobson & Dove, 1975) which may enable us to supplement work on motility in plasmodia, if the same regulatory genes are involved in both states of *Physarum*. If not, we may be able to characterise genes that are selectively controlled in two distinct life forms.

This brings us directly to the third and, up to now, main field of interest in *Physarum*: growth and differentiation, and to the McArdle Laboratory for Cancer Research in Madison, Wisconsin. There H. P. Rusch has adopted *Physarum* as a model system and he has worked on it in cooperation with numerous postdoctoral fellows.

Although synchronous nuclear divisions of the plasmodium have been known since their discovery by Howard (1932), *Physarum* was rediscovered by a professor of human oncology in a search for a simple experimental system to study basic mechanisms of growth and differentiation, and to find a cure for cancer (Rusch, 1980). Rusch was given a specimen of *Physarum* by Myron Backus, a mycologist on the Madison campus. He became convinced that the plasmodium does either grow, or differentiate, and would provide sufficient material for a biochemical analysis of two alternate states of development, if he could succeed in growing the organism in pure culture and induce sporulation at will.

Rusch's proposal to tackle the cancer problem in this way may have been unorthodox at the time he began to work on *Physarum*, but is quite usual today. He looked at the cancer cell not as one that has lost control over its proliferation, but as a cell that has failed to differentiate properly. Hence, his approach has been to extract an active principle from, say, a sporulating culture of *Physarum*, and apply it to a growing plasmodium, thereby forcing it to differentiate and abandon the proliferative mitotic cycles for ever. This aim has not been achieved yet, but we hope, in discussing the relevant observations, that *Physarum* is a true model to analyse differentiation phenomena in every respect that interests the developmental biologists. In the McArdle Laboratory the groundwork was laid, and an explosion of biochemical data that are conceptionally held together by their possible relevance to the cell cycle has taken place. Even though the plasmodium is not a proper cell, more information on the lives of a cell may be gleaned from *Physarum* than for any other cell.

We shall now trace some of the people and their work and include essential papers and reviews by the respective researchers. In the fifties, J. W. Daniel joined Rusch. It took them about a decade of trial and error for a liquid medium to be concocted in which *Physarum* would grow in the absence of other microorganisms, i.e. axenically (in the presence of glucose and hematine), or sporulate in the absence of nutrients but in the presence of niacin (Daniel & Rusch, 1961, 1962).

The Gutteses, a dedicated husband and wife team, carefully described the timing and morphological change in the nuclear division cycles and sporulation (Guttes, Guttes & Rusch, 1961). E. Guttes & S. Guttes (1964) also developed the procedure to fuse many microplasmodia into one macroplasmodium and made numerous types of fusion experiments with macroplasmodia in a quest for the regulatory mechanisms controlling nuclear division (1968).

W. Sachsenmaier carried out fusion experiments (Rusch *et al.*, 1966) and discovered fluctuations in the activity of thymidine kinase, which coincided with the nuclear division cycle (Sachsenmaier, 1976). He is the main proponent of the 'hour glass' model of mitotic regulation (Tyson & Sachsenmaier, 1979) and has been searching for a cytoplasmic triggering substance.

O. F. Nygaard has made the first analysis of DNA synthesis and discovered that DNA replication sets in immediately after mitosis, leaving no G_1 phase for this naturally synchronous cell cycle (Nygaard, Guttes & Rusch, 1960). In a classical paper it has been shown by an elegant density labelling experiment utilising a heavy DNA precursor (bromodeoxyuridine, BUDR) instead of thymidine and two radioisotopes (^{3}H and ^{14}C) that DNA replication follows a strictly sequential pattern in consecutive cell cycles (Braun, Mittermayer & Rusch, 1965). Incorporation of [^{3}H]uridine indicated a biphasic RNA synthesis in the cell cycle with one activity peak in S phase and the other in G_2 phase. A first qualitative characterisation of RNA was attempted by J. E. Cummins (Cummins, Weisfeld & Rusch, 1966). He also was coauthor of a review summarising the beauty of the system and displaying some colour photographs from an instructive film by the Gutteses and Rusch (Cummins & Rusch, 1968) which has been shown by Rusch along with the accumulation of experimental and chemical data at many international conferences and which initiated interest in *Physarum* in many laboratories.

Joyce Mohberg devised a simple method to isolate purified nuclei (Mohberg & Rusch, 1971), and although they do not as yet divide in the test tube, they have retained some of the regulatory factors operating *in vivo*, as they continue to make DNA in S phase only and display the biphasic pattern of RNA synthesis. RNA transcription has been most extensively analysed by R. Braun (Seebeck & Braun, 1980) and Melera & Rusch (1973). P. W. Melera has also characterised the tRNA (Melera, Momeni & Rusch, 1974) and identified mRNA in *Physarum* (Melera, Davide & Hession, 1979). DNA synthesis *in vitro* has been studied by Brewer & Rusch (1966) and by Schiebel & Schneck (1974). Mohberg also measured the DNA content, succeeded in counting chromosomes in isolated metaphase nuclei, and has become the authority in this field (Mohberg, 1974). Furthermore, she has done a thorough analysis of the histone composition of *Physarum* (Mohberg & Rusch, 1970). In addition, together with the expert technical staff on the seventh floor of the McArdle Laboratory, she was at the heart of the Rusch group for many years and freely gave away her tricks about how to handle *Physarum* to many generations of postdoctoral fellows. She has been diligently dispatching spherules of *Physarum* to interested people all over the world. She has also spent several years in European laboratories where her expertise in setting up cultures has helped progress in life-cycle analysis in Leicester, England and in the control of the cell cycle in Innsbruck, Austria. Finally, she has kept track of the various strains of *Physarum* in use and compiled a detailed genealogy (with Kay Babcock) first communicated in a *Physarum Newsletter* (Babcock & Mohberg, 1975).

The work on nuclear proteins was continued by Brigitte Jockusch (1973), and detailed analyses of non-histone proteins have been initiated by W. M. LeStourgeon (LeStourgeon *et al.*, 1978) and has been carried on since (Jeter & Cameron, 1974), while those proteins with an affinity for DNA have been analysed by Magun (1976).

E. M. Goodman had begun work on *Physarum* with the electron microscope before he detected glycogen particles and polyphosphate at the McArdle Laboratory. He has since studied the effect of very weak, low-frequency electromagnetic fields on the cell cycle and has just published a comprehensive up-to-date review on *Physarum* (Goodman, 1980), concentrating on structural-functional aspects. Together with W. Schiebel he has isolated cytoplasmic fractions and tried, unsuccessfully, to establish an assay for the elusive mitotic factor. Schiebel has since turned to DNA polymerase work and has written a concise critical review on the cell cycle (Schiebel, 1973).

J. McCormick analysed the composition of slime and the wall material, and has asked which enzymes are involved and whether these are just activated or newly synthesized (McCormick, Blomquist & Rusch, 1970). A. Hüttermann has been interested in the metabolic changes during spherulation, and he has introduced density labelling of *Physarum* proteins by heavy amino acids, to analyse de-novo enzyme synthesis (Hüttermann, 1973) and explained that the massive slime production during starvation is due to gluconeogenesis. I. Chet has analysed the RNA metabolism during spherulation and introduced the mannitol technique in an attempt to separate phenomena of starvation from those of differentiation (Chet & Rusch, 1969). W. D. Grant (1972) began his work with isolated nuclei and provided the first evidence for multiple RNA polymerases typical for eukaryotic cells. He has also been involved subsequently in theoretical models of the cell cycle (Sudbery & Grant, 1975).

A role for light in sporulation had been suspected since early on (Baranetzki, 1876) and was demonstrated by Gray (1938). The irreversibility of this process was studied by Ward (1959) in impure culture, and the work of Daniel (1966) and the Gutteses (Guttes, Guttes & Rusch, 1961) had clearly outlined the time sequence of morphogenetic, cytological and physiological events. A first attempt to understand the differentiation process at the level of gene regulation has been made (Sauer, Babcock & Rusch, 1969*a*). This work has since been extended to spherulation and the cell cycle (Sauer, 1973, for review).

If this brief listing of the many postdoctoral fellows sounds like a medley of independent projects, the results of this work done during the sixties have been coherently presented by Rusch on two occasions (Rusch 1969, 1970).

The following decade saw work on growth and differentiation of *Physarum* at various places with some old and many new people, and a constant intermingling has taken place among researchers and the strains of *Physarum*. The results can be deduced from an authoritative monograph edited by Dove & Rusch (1980) with a complete list of references.

One synthetic approach of the whole field has begun with the first paper on *Physarum* genetics (Dee, 1960). Genetical techniques have been systematically employed, or had to be developed by Dee (1975). F. B. Haugli initiated mutagenesis work on amoebae with Dove at the McArdle Laboratory (Haugli, Dove & Jimenez, 1972). He has set up a highly efficient laboratory at Tromsø, continuing his search for meaningful mutants, and performing in parallel a series of experiments on DNA replication (Funderud, Andreassen & Haugli, 1978), in an effort to combine both lines of work. Work on DNA synthesis has been continued (Brewer & Ting, 1975), and the early replicated DNA has been isolated by S. Shall (Beach, Piper & Shall, 1980). At the McArdle Laboratory mutant work continues with a particular thrust to describe developmental mutants of the amoeba–plasmodium transition (Gorman, Dove & Shaibe, 1979). Such is the main line of experiments in the productive laboratory of C. E. Holt, where many important developmental mutants have been found (Adler & Holt, 1977). In addition, in collaboration with Hüttermann and the Institut für den wissenschaftlichen Film in Göttingen, Germany, a series of films giving a full account of *Physarum* as model for cell biology studies has been completed.

Some progress has been made in the photobiology of sporulation (Rakoczy, 1980), and preliminary injection experiments with extracts (Wormington, Cho & Weaver, 1975).

A main centre for molecular biology of *Physarum* has been established by Braun in Bern, Switzerland. A detailed analysis of the fine structure of rDNA isolated together with an intact extrachromosomal mini-chromosome, plus the extensive purification of RNA polymerases and the characterisation of nucleosomal structure of the chromatin are examples of its work (Seebeck & Braun, 1980, for review). The first paper involving recombinant DNA techniques stems also from this laboratory (Gubler, Wyler, Seebeck & Braun, 1980). RNA transcription has been further analysed *in vivo* and *in vitro* (Davies & Walker, 1977; Sauer, 1978; Turnock, 1980), and structural-functional relations of chromatin are being extensively studied in several laboratories (Bradbury *et al.*, 1974; Grainger & Ogle, 1978; Johnson, Campbell & Allfrey, 1979; Jerzmanowski *et al.*, 1979). An important hypothesis is that histone phosphorylation could be a decisive event that triggers mitosis, and initial experiments with exogeneous histone H_1 kinase have supported this notion (Matthews & Bradbury, 1978). Recent analysis of the sequence organisation of the DNA confirms the idea that *Physarum* is organised more like higher animals than unicellular organisms (Hardman & Gillespie, 1980; Hardman *et al.*, 1980).

All current activities on *Physarum* are being put together in an ambitious treatise edited by Aldrich & Daniel (1981) containing over 25 individual chapters, whose authors read like a 'Who's Who' in *Physarum* research, combining the main areas of research: life cycle, motility, growth, differentiation, genetics, and a most valuable appendix on techniques. It will not be long before the reader can learn everything he always wanted to know about *Physarum*. As I have been preparing for this manuscript for some time, and having been in touch with the editors and most contributors, I am aware that much can be said about *Physarum*, yet little attention has been focussed on the problems and possible progress in questions of developmental biology which may be extracted from this unique system. Developmental biology is the theme of this book. Some interrogations with *Physarum*, in the laboratory and from the literature, with respect to development have been presented in the first descriptive sections. Later on we shall come to some selected experiments, but first we need to take a plasmodium apart and assemble a few essential facts about the pieces.

Before we do that, and after having introduced so many Physarologists, a few more words about *Physarum polycephalum*. Most cultures that are carefully kept in the liquid medium on a shaker in the dark room (in constant fear of contamination, and of ever more clever and, in most cases, more expensive experiments) are derived from the specimen collected by Backus and given to Rusch. This strain is now known as Wis 1. From its sclerotium, two series of sublines were started at the McArdle Laboratory. The first series M was adapted to growth in axenic medium in about 1959 and three sublines, M_{3a}, M_{3b}, and M_{3c}, were established from it. One specimen of M_{3a} was shipped to Leicester, where the genetic work was begun about 1959, it was also the source of the incompatibility strains studied first by Carlile & Dee (1967). While these lines were all originated from vegetative macrocysts, new sublines were created from M_{3c} by mating of amoebae, and several sublines M_{3c} IV–VIII are widely used.

The second series was also started from the Wis 1 sclerotium in about 1965, and amoebae were isolated and grown on bacterial lawns by a graduate student R. Steeper, hence named RS. These were used in the original attempts at mutagenisation by Haugli. They were also the source for the first amoebae RS D4 to be adapted to liquid medium by Goodman (1972), and they were mated to produce plasmodia. RS amoebae were also taken to Tromsø University and renamed TU for a series of sublines created in the genetic work there. Further lines were selected by Jacobson and Dove (hence DJ) which readily can be taken through the whole life cycle.

A second strain, Wis 2, was isolated in the field again by Backus about 1957 and was shown to be of a different mating type from Wis 1 by Dee (1973). A different strain was obtained from the Iowa State University Farms and became the source of many life-cycle studies and mutant strains (Collins, 1979). A strain from Indiana was the source for the morphological and taxonomic work of Alexopoulos and the extensive genetical analysis at the Massachusetts Institute of Technology by Holt, hence named the CH-series.

A very important strain, Colonia (C) was obtained by von Stosch from the botanical garden of Cologne University. This apogamic strain (which may have originally come from the Wisconsin isolate), was adapted for routine genetic analysis in Leicester (hence Cl), and this derivative is now used extensively in mutagenesis work. Yet another strain was worked on by Kamiya and other Japanese researchers.

One very useful mutant is a white form of *Physarum* (Anderson, 1977). As this plasmodium does not fuse with the M series, they can be used to demonstrate incompatibility of plasmodia. One can also ask whether the yellow pigment is required for sporulation (it is not), and can monitor amoebal crosses and discriminate them from plasmodial fusions with the naked eye (provided the right strains are used).

This brief outline shows the intricate relationships between strains and sublines of *Physarum* and is based on the newsletter report by Babcock & Mohberg (1975). One may wonder why people have not concentrated on one single line of *Physarum*. The answer is that work on this organism has evolved from several places, and there are subtle differences between the strains: some grow well, but are poor in spherule formation; others are poor in sporulation incidence or germination of spores, others age rapidly, and in some strains mitosis can be better predicted and is more synchronous. Furthermore, the famous strain Cl does not grow in perfectly balanced condition.

Nevertheless, as most biochemical work has been done on the Wis 1 isolate in the McArdle strain and most mutagenesis and genetic analysis is now being done on the Cl strain, and since hard facts will only emerge if both approaches are combined, Physarologists should be urged to concentrate on one strain, even if it calls for some duplication of work already published. Such forthcoming results will also show whether some of the general concepts – like a biphasic pattern of RNA transcription in the cell cycle, to name but one – can be confirmed in a different strain of *Physarum* and bring us on common and more solid ground.

3

Dissecting the plasmodium

Many questions about the life cycle of an organism can only be answered after it has been killed – gently. Subcellular structural aspects can be well described with the electron microscope, and an insight into functional aspects is often gained after grinding up the cells and dividing the homogenate into subfractions by various centrifugation procedures. A number of constituents of a plasmodium are shown in Fig. 15. The main subcellular organelles, nuclei and mitochondria, can be clearly seen along with various vesicles and small granules, most of the latter consisting of glycogen, *Physarum*'s energy store. There is no elaborate endoplasmic reticulum, nor do any microtubules form a rigid scaffold, probably because of the constant streaming. But most remarkably there are no cell membranes separating the nuclei from one another. This latter feature, together with the fact that there is no rigid wall around the whole plasmodium, allows for very gentle homogenisation in a blender.

If a plasmodium is suspended in a buffer containing a detergent (like Triton X-100), sucrose (to avoid clumping or swelling of the nuclei) and calcium ions (to stabilise the nuclear envelope), the nuclei can be almost quantitatively recovered by centrifugation. If the homogenate is layered over a cushion of 1 M sucrose, only the nuclei will sediment as a pellet, and the other particles will stay in the supernate. Variations of this procedure, originally devised by Mohberg & Rusch (1971), have yielded good preparations of mitochondria and ribosomes. There is a large amount of fibrils on the surface of the plasmodium. This extracellular material, the slime, is probably very useful to the slime mould in protection and migration, but has been a nuisance for biochemists in preparing clean extracts. This polysaccharide consists of galactose and contains sulphate and some phosphate groups, it cannot be metabolised by the plasmodium, and no enzyme has been found in any of many microorganisms tested which degrade this polymer. Nor is a slimeless mutant available. However, myxomycete amoebae do not produce slime, or produce just enough to secrete a thin coat (glycocalyx) to cover their bodies.

A unique method of preparing pure cytoplasm involves just spinning a macroplasmodium in the ultracentrifuge. This method allows for a clear-cut localisation of some molecules. For instance, the question as to whether the

Fig. 15. Components of the plasmodium. The plasma membrane (1) is highly involuted, its surface is covered with slime (2); the nucleus contains a nucleolus (3), shows the typical envelope (4) with pores and chromocentres (5). The cytoplasm contains prominent glycogen granules (6) and ribosomes (7); mitochondria (8), pigment granules (9), food vacuoles (10), contractile vacuoles (11) and pinocytotic vesicles (12), lysosomes (13), dictyosomes (14) are present among the regular constituents.

RNA polymerases are in the nucleus has been answered affirmatively, even though only a small fraction of those enzymes are found inside the nucleus after its isolation in the aqueous medium mentioned above.

We shall now summarise some observations of subcellular components that will help us to enquire into the processes of growth and differentiation. This field has been reviewed recently by Holt (1980) and Goodman (1980), where most references can be found.

The nucleus

Isolated diploid nuclei of *Physarum* are spheres of about 5 μm in diameter. As the nuclear membrane is present at all stages of the nuclear cycle, nuclei can even be prepared in metaphase complete with mitotic spindle plus chromosomes, about 80, and maybe someone will succeed in making those nuclei divide, either after injecting into frog oocytes or in the test tube. Nevertheless these nuclei retain some of their vital functions, and several research groups have analysed in-vitro DNA synthesis (replication) and RNA synthesis (transcription). We will come to that aspect later on, after a look at the structural composition.

The amount of DNA per nucleus has been determined as about 1 pg (10^{-12} g), which seems very little but it represents a thread of double-stranded DNA of 30 cm – a lot for a nucleus of only a few microns in diameter. Several methods have been used. In one the DNA is specifically stained inside the nucleus by the Feulgen reaction, and its quantity is determined in a spectrophotometer for a large number of individual nuclei. As the variability among the staining of single nuclei is very low, another method, that of extracting DNA from a large number of nuclei for biochemical quantification, gives equally reliable results. A third method, which allows averaging of the DNA content in thousands of nuclei with the help of a computerised flow cytophotometer, reveals a symmetrical distribution which proves that we are dealing with a homogeneous population of most nuclei. If we compare the amount of DNA immediately after nuclear division with that measured 5 h later, in G_2 phase, we can see that the amount of DNA per nucleus has doubled, as one would expect. Therefore, the amount of DNA in G_2 phase (1 pg), in a diploid nucleus with 80 chromosomes, is four times greater than in the unreplicated haploid genome. As a result of these measurements we find 0.25 pg DNA per genome or about 85 mm. This much DNA in *Physarum* is approximately a hundred times more than that of a bacterium, ten times more than in *Dictyostelium*, and only about ten times less than in higher organisms like man. For one chromosome of *Physarum*, this amounts to 2 mm of DNA, and as a metaphase chromosome is less than 1 μm long, the DNA has to be shortened at least 1000 fold, just as in other eukaryotic chromosomes.

The length of DNA can be easily translated into molecular weight or number of base pairs (bp) by the one-two-three rule, which says that 1 μm double-stranded (ds) DNA equals 2×10^6 daltons, for 3000 bp. Very gentle extraction procedures have yielded DNA molecules up to 1000 μm in estimated length, which could indicate that a chromosome contains a single DNA molecule. However, if the dsDNA is made single-stranded (ss), a much smaller size (not longer than 10 μm) per ssDNA molecule is obtained. This could indicate that, at regular intervals, there may be single-stranded breaks in the DNA molecules. These may have occurred during the preparation, or they may be present in the native DNA molecules and be closed by 'linkers', perhaps protein molecules. This is an important and long-standing problem in DNA research in higher cells and is currently debated in mammalian DNA once again (Krauth & Werner, 1979).

Two other general properties of *Physarum* DNA are the melting temperature (85.8 °C) and the buoyant density (1.700 g/cm^3) which allow an estimate of the base composition of the DNA; direct measurements have confirmed the following values for the chromosomal DNA: 28% A, 30% T, 22% G, and 20% C. If the total DNA of a plasmodium is analysed on a caesium chloride gradient, three bands can be distinguished, a main band, a heavy satellite (about 2%) and a light satellite (about 7%). The heavy one is in the nucleus and is located in the nucleolus, the light one is derived from mitochondria, and both will be described below.

The main band, or chromosomal DNA, has been analysed with respect to its sequence organisation. It is known that eukaryotic organisms contain DNA sequences which are only present once (or a few times) in the genome, the unique DNA (=single-copy DNA), and others that are present in many copies, the repetitive DNA sequences. This is also true for *Physarum*, and three points can be made: (i) about 5% of the denatured DNA reassociates immediately, due to intramolecular association of so-called inverted repeats; (ii) about 40% is made up of repetitive sequences; and (iii) the rest is unique DNA. The sequence complexity of the latter is estimated as 1.7×10^8 bp (for comparison, *E. coli* has 4.5×10^6 bp and *Dictyostelium* 5.4×10^7 bp). If we assume that all this unique DNA contains genetic information, and a mean length of mRNA of 1500 nucleotides, *Physarum* could code for 10^5 different proteins, which is a respectable number. A simplified version of more reassociation data and ultrastructural characterisation is summarised in Fig. 16. The inverted repeats, after denaturation, can form base-paired 'hair pins', unless they are separated by a piece of unique DNA (then they are part of a tennis-racket structure). The repetitive sequences are interspersed with unique sequences, and on average there is one inverted repeat every 7000 nucleotides, and the repeated and unique sequences measure 600 and 1000 nucleotides, respectively. The repetitive sequences belong to about 100 different groups (families); each family has about 2000 members. Although there is great variation in these measurements, the arrangement is clearly non-random and, on the whole, quite similar to other eukaryotic organisms. So far no sizable fraction of short, tandemly arranged, highly repetitive DNA (so-called satellite DNA) has been described.

Nuclei contain about five times as much protein as DNA. Two-thirds of the protein belongs to the nucleolus. One prominent class of proteins, which are present in an amount equal to that of the DNA, are the histones. As these proteins are highly conserved in evolution, similarities with calf histones in their pattern after electrophoresis can be seen (Fig. 17). The slow-moving protein which is rich in lysine has been classified as H_1. Of the other four histones, two migrate in a similar way to H_3 and H_4 of calf; the

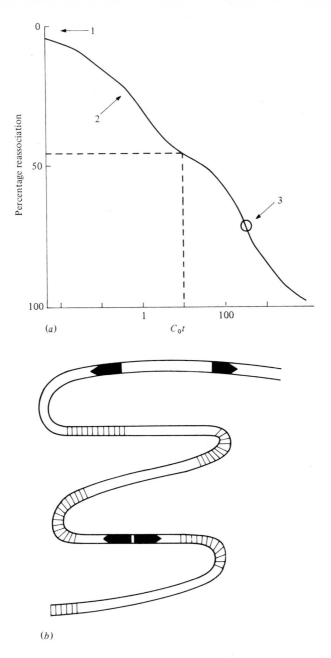

Fig. 16. Organisation of the chromosomal DNA. (*a*) The kinetics of DNA reassociation (C_0t-analysis) reveals three components: 1, inverted-repeat sequences 5%; 2, repeated sequences 40%, and 3, unique sequences *c.* 55%; its C_0t ½ value (approx. 500) allows an estimate of the genome complexity (semi-schematic). (*b*) Schematic depiction of the rather regular interspersion of repeated (hatched) and unique DNA sequences, and the inverted repeats (arrows).

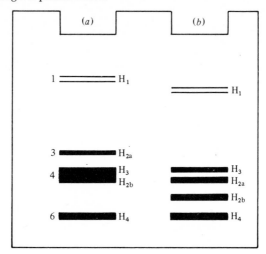

Fig. 17. Schematic representation of the nuclear histones of *Physarum* (*a*) as compared to calf histone (*b*) patterns, and tentative assignment of bands (1–6) and histone species (H_1–H_4) after SDS-polyacrylamide gel electrophoresis.

remaining two differ somewhat in mobility from H_{2a} and H_{2b} of calf thymus. Nevertheless, the basic organisation of chromatin in *Physarum* resembles that of any other eukaryotic cell and can be best described as beads on a string. The beads have been analysed on gels or sucrose gradients after nuclease digestion. Each one contains two molecules of each histone (except H_1), and a stretch of DNA of about 200 bp length is wound around the bead; thereby the DNA becomes shortened about seven-fold (see Fig. 19).

From the amount of DNA per nucleus, the number of beads and consequently of histone molecules can be calculated to be about 10^{10} per nucleus. This number reveals a real problem if one wants to find any regulatory protein molecules in the nucleus. Those are estimated to occur at a concentration of 10^4 molecules per nucleus and require methods other than descriptive biochemistry to be detected. Nevertheless, the patterns of nucleosomes have revealed a fraction (peak A), that moves slower on sucrose gradients and contains a deficit of histone, yet some additional proteins of the so-called high mobility group (HMG) which are suspected to indicate active genome regions. This observation together with the high sensitivity of the active chromatin to DNase I digestion and a smooth fibrillar structure in the electron microscope may allow us to differentiate inactive from active chromatin on a merely structural basis, the latter amounting to $15 \pm 5\%$ of the DNA.

Another highly abundant protein component has been isolated from nuclei that have been extensively digested with DNase to solubilise the chromatin. The residual insoluble structure has been called nuclear matrix and consists mainly of two proteins of about 50 000 and 25 000 daltons. It may be significant for the organisation of the chromatin in interphase that the residual DNA (less than 5%) in these preparations contains a high proportion of freshly replicated DNA (Mitchelson, Bekers & Wanka, 1979).

Yet another fraction of protein, which belongs to the chromatin, comprises those that are complexed with RNA, particularly with nascent RNA (that which is freshly synthesized). Although distinct ribonucleoprotein (RNP) particles, like the 40 S particles in mammalian cells, have not been found in *Physarum*, at least one of the RNP proteins seems to be highly conserved. It contains the same unusual amino acid (N,N-dimethyl-arginine), and its antibody cross-reacts with mammalian RNP particles. It has even been claimed to react to the puffs of giant chromosomes of *Drosophila* (Walker *et al.*, 1980).

Two other characteristic proteins have been demonstrated in clean nuclear preparations: actin and myosin. Furthermore, comparing interphase with metaphase nuclei, tubulin has been found as a dominant fraction in the latter.

In addition to the proteins identified so far, there is a large number of protein bands on gels that have been noted consistently, like a group of non-histone proteins (NHP) that can be solubilised in phenol. Definite changes in the composition have been demonstrated by electrophoresis which can serve as landmarks at the molecular level for certain physiological states (like starvation) of *Physarum*, where no cytological markers are available.

Finally, a class of proteins has been described in the soluble fraction of the whole plasmodium that selectively bind to ss- or dsDNA of *Physarum*. Among these are a number that are phosphorylated. It may be significant, that these phosphoproteins disappear from a plasmodium with age (Magun, 1974).

Two kinds of enzymes which bind strongly to DNA, DNA-, and RNA-polymerases have been extensively studied. Again, as with all proteins studied to date, they seem to be abundant molecules of the nucleus. Even at maximum activity, there are many free enzyme molecules of these polymerases which seem not to be engaged in the polymerisation processes.

Another major macromolecule which is definitely present in substantial amount in the nucleus of *Physarum*, about 1.5 times that of the DNA, is a polysaccharide, a β-galactan, which is somewhat different from the slime.

The nuclelolus

Nucleoli have been isolated, either from whole plasmodia or from isolated nuclei. Although they contain, to a variable degree, all the protein components mentioned for the whole nuclei, they are the exclusive site of ribosomal DNA. This may turn out to be an over-statement, but until now nobody has been able to demonstrate a gene for rRNA in the chromosomes. This is a unique situation for any eukaryotic cell, where the nucleolus organizer is an important part of at least one chromosome per genome. Aside from this negative conclusion, the organisation of the nucleolus is quite remarkable: it is an aggregate of some 300 mini-chromosomes, each containing a piece of DNA complete with the essential requirements for autonomous replication and transcription of a single gene, a gene whose product is superabundant and well characterised: the rRNA that is associated with protein in the ribosomes. This situation is different from the amplification of rDNA in many oocytes, which is a temporary phenomenon (Davidson, 1976). Structural similarities to the rDNA exist in the macro-nucleus of some ciliates, but in *Tetrahymena* one copy of rDNA has been located inside the micronucleus from which the macronucleus originates (Gall, 1974). At any rate, the ease with which quantities of rDNA can be isolated, has made the nucleolus of *Physarum* an interesting model to study chromatin structure and function by restriction analysis, and molecular cloning has been successfully used for fine structure analysis (Fig. 18).

The nucleolar DNA is denser (1.714 g/cm^3) than the chromosomal DNA and amounts to 2% of the total DNA. Most molecules are linear, 20 μm long (39×10^6 daltons). Upon renaturation the ssDNA molecule does not renature with another one but with itself, i.e. in an intramolecular fashion. There is a symmetry axis just in the middle, making the whole piece of DNA a big palindrome or an inverted repeat of 10 μm length. Consequently, each base sequence along the 20 μm stretch of DNA occurs twice, hence two genes are to be expected on each piece of DNA. Two such genes for rRNA have been located at the ends; each is 4.2 μm long, which leaves a very long spacer region in between of about 10 μm. This stretch contains short, repeated sequences as deduced from renaturation pictures. In a chromatin-spread picture the spacer has been measured to be 6.2 μm. Nuclease digestion and histone composition indicate differences in the transcribed, i.e. active, chromatin, supporting the idea of definitive structural differences in active chromatin as mentioned above for the whole chromatin. A series of nucleolar DNA molecules with 'eyes' has been interpreted as evidence for a replication origin close to the middle of the molecule. Since RNA polymerase A is also an integral part of the nucleolar mini-chromosome, it can be anticipated that an accurate knowledge of structure and function of the rDNA-replicon of *Physarum* will soon be available (Seebeck & Braun, 1980).

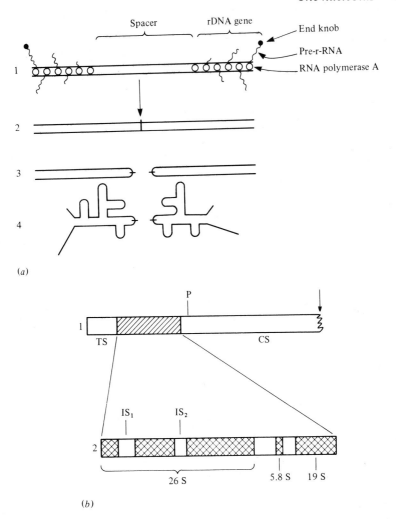

Fig. 18. Organisation of the ribosomal genes in *Physarum*. (*a*) Nucleolar chromatin has been claimed to be a chromatin fibril with two matrices of nascent ribonucleoprotein fibrils (4.2 μm each), separated by a large spacer (11.6 μm) (1). Purified nucleolar DNA contains linear molecules of 20 μm double-stranded DNA (2); after denaturation each strand can re-anneal around the rotational axis of the palindrome (arrow) and yield one large hairpin structure (3) or shorter inverted repeats in the spacer region (4). (*b*) One half of the rDNA-palindrome (1) displays the axis of rotation (arrow), the central spacer (CS), and the terminal spacer (TS), and the rDNA gene which is transcribed starting at the putative promotor site (P). Detailed analysis of the transcribed gene (2) has revealed the location of the respective DNA sections for 19 S rRNA, a 5.8 S rRNA, and of the 26 S rRNA which contains two introns.

Mitochondria

Numerous oblong particles can be seen in the plasmodium. They are about 0.2 μm thick and variable in length (0.3–2.5 μm). These mitochondria can be isolated intact, so that respiration can be studied *in vitro* (Holmes & Stewart, 1979). While their main function in the intact organism is energy production, some observations may be worthwhile for our purposes. First, the inner membrane is made up of a large number of cristae. They are of the tubulus type, which is found in protozoa, and are not organised in the form of lamellae as in higher eukaryotes. Second, the various shapes of the organelles strongly suggest that they multiply by dividing in two after a growth period. Third, within each mitochondrion lies a rod-like structure, the nucleoid, which contains DNA and at least one histone-like basic protein of 32 000 daltons. This DNA has been isolated: it amounts to 5–10% of the total DNA, and its density is 1.686 g/cm^3 which is less than that of the chromosomal DNA. The molecular mass of 40×10^6 daltons is almost equal to that of the nucleolar DNA. Most molecules (90%) have been identified as linear structures in the electron microscope, but DNA circles have also been described which may be held together by a protein (Bohnert, 1977). An attachment site of the DNA complex at the outer membrane of the mitochondrion has been postulated from autoradiographic studies, which is believed to control separation of equal numbers of genomes after division. Approximately 40 molecules of DNA have been estimated per nucleoid. This is a high value compared with higher cells in both molecular size and in amount. One interesting aspect lies in the question of whether mitochondrial DNA replication and division are in phase within the replication programme of the nucleus.

Mitochondria contain ribosomes of somewhat smaller size than in the cytoplasm. They seem to contain an endogeneous RNA polymerase which is sensitive to rifampicin and produce some RNA that has not been further studied. As a final comment, these mitochondria can accumulate calcium ions *in vitro* and, particularly after illumination of the plasmodium, *in vivo*.

Ribosomes

The ribosomes of *Physarum* measure 14 nm in diameter, and the monosome sediments at 80 S which is the typical value for eukaryotic cells. The large subunit contains the large rRNA of 26 S, and the small subunit the small rRNA of 19 S. In addition to the large RNAs a 5 S rRNA and the 5.8 S rRNA have been characterised. The former is transcribed from about 1000 genes of the chromosomal DNA, the latter is part of the nucleolar rDNA transcript.

The proteins of the subunits have a similar complexity to those of other higher cells. It may be important for the regulation of protein synthesis that three proteins occur in a phosphorylated form. One of them may be the same as in rat liver (Bélanger, Bellemare & Lemieux, 1979). Ribosomes have been detected at the outer nuclear membrane. Otherwise, at least during growth, they seem not to be membrane-bound, as typical endoplasmic reticulum (ER) is absent. Polysomes in the process of protein synthesis have been extracted and displayed on sucrose gradients. Reproducible patterns of large polysomes have been best shown from microplasmodia (Turnock, 1980), whereas from other stages good preparations have not yet been obtained consistently. A recent improvement may ease the uncertainty in the characterisation of cytoplasmic RNA, particularly mRNA (Adams, Noonan & Jeffery, 1980).

Other constituents

There are numerous vacuoles in the plasmodium – like food vacuoles, contractile vacuoles, pinocytotic vesicles, lysosomes, and coated vesicles – which may all be part of an elaborate mechanism of membrane flow. The Golgi apparatus is poorly developed, appearing as a stack of oblong vesicles in the growing plasmodium. In addition there are vesicles which can accumulate calcium in an ATP-dependent reaction, and abundant pigment granules (possibly plant-type spherosomes) which contain the typical yellow pigment. Glycogen granules of 20 nm serve as an energy store.

The complex mixture of soluble material that is left after a high speed centrifugation contains monomeric forms of actin and myosin, the main structural proteins in *Physarum*, and all other ingredients of the cytosol, many enzymes, tRNA and pools of nucleotides, and amino acids. Many other metabolic substrates have been identified under various conditions of plasmodial life.

4

Molecular biology of *Physarum*

The previous chapter has stressed the structural organisation of the nuclear genome. Before we can turn to the more relevant question of the regulation of genome expression in the life cycle of *Physarum* we first need some information on the general functional aspects of DNA replication and RNA transcription, and also of protein synthesis (translation). Again, most of the original results have been reviewed (Hüttermann, 1973; Braun *et al.*, 1977; Holt, 1980; Seebeck & Braun, 1980; Turnock, 1980; Walker *et al.*, 1980).

Replication of chromosomal DNA

DNA synthesis can be conveniently studied by radioactive thymidine incorporation. It takes about 2.5 h to replicate the chromosomal DNA, irrespective of the chromosome number or of ploidy. DNA replication should be semiconservative, and this has been confirmed by density-labelling with bromodeoxyuridine (BUDR): after one S phase, practically all of the DNA becomes heavy, and after the following S phase in BUDR, half of the DNA stays heavy and the other half gets heavier still, as would be expected for heavy–light (HL) and heavy–heavy (HH) molecules. As in all eukaryotes, genome replication occurs in many places at one time, i.e. discontinuously. This has been seen in fibre autoradiographs (Funderud, Andreassen & Haugli, 1979), and deduced from electron microscope pictures. Each stretch of DNA that is thus replicated has been called a replicon; in *Physarum* it is 10–20 μm long. Within each replicon, replication starts at one site, the origin. An elegant experiment suggested that in *Physarum* replication is bidirectional, as in many other eukaryotes. For a short period, at the beginning of S phase, bromodeoxyuridine (BUDR) was incorporated into replicating DNA (instead of the usual thymidine) to make the newly synthesised DNA strands sensitive to UV irradiation. These strands were also labelled with [³H]deoxyadenosine to allow their detection later on. Then DNA replication was allowed to continue for one hour in the absence of BUDR. DNA was isolated and irradiated with UV light (320 nm) to break the strands at those points where BUDR had been incorporated. The size distribution of irradiated and control DNA was analysed on alkaline sucrose gradients. As all the ³H-labelled DNA strands were photolysed exactly in

56

the middle, they must have been started at the same time at one place within a replicon, at the origin, and grown at the two replication forks simultaneously and with the same speed.

Autoradiographic tracks show that DNA synthesis proceeds at a speed of 2–4 μm in 5 min. Brief labelling periods and sizing of the newly labelled DNA by alkaline sucrose gradients or agarose gels have shown that at first the new DNA occurs in small pieces, so-called Okazaki pieces, each about 250–300 nucleotides long. These become joined, about four each minute, at a speed of 0.2 μm/min, until the full size of a replicon has been reached. This process of joining is quite complex, since an Okazaki piece of DNA begins with a short piece of RNA. It has been definitely shown with *Physarum* that a covalent linkage between any of the four ribonucleotides with any of the four deoxyribonucleotides does exist (although those beginning with dG are more frequent – Waqar & Huberman, 1975). This requires that the short piece of RNA has to be removed enzymatically, leaving a gap which in turn has to be filled with DNA. Only then can the ligation of two Okazaki pieces take place.

It is not yet clear whether such short pieces occur in both daughter strands. Theoretically one strand, the leading strand, could be synthesised continuously, and the observed small pieces in that strand could result from faulty insertion of a wrong nucleotide and could require DNA repair. At any rate, a replicon of about 20 μm would be completely replicated in 100 minutes if only one replication fork were moving, and half as long if replication were bidirectional, a situation that seems to prevail in *Physarum*.

Up to now, the growth of the new DNA strand, despite the complex mechanism just mentioned, can be called continuous. Now comes a pause. A stepwise growth of the chain will occur joining about four to five replicons (a cluster), and after another pause four to five of such clusters will join together. This leads to very long DNA molecules of about 1 mm and takes about two hours.

The Okazaki pieces are about equal in length to the DNA stretch per nucleosome. The cluster of replicons (50 μm) and the cluster of replicon-clusters (250 μm) may reflect a higher order of chromatin structure, involving scaffold proteins, the nuclear matrix and the nuclear pore complexes.

Some progress has been made in developing an in-vitro system of DNA replication. It requires intact S-phase nuclei, all four dNTPs, ATP, and is sensitive to detergent. Homogenates suspended with dextran or a cytoplasmic glycoprotein of 32 000 daltons increase the efficency of the system, which can reach 15% of net synthesis of the genomic mass at 25% of the in-vivo rate, in a process in which pieces similar to the size of Okazaki pieces are joined in part to replicon-size DNA (Brewer & Busacca, 1979). An elegant test has been performed to assure that the DNA made *in vitro* represents DNA replication and not DNA repair-synthesis. Nascent DNA was density-labelled with BUDR *in vivo*, then nuclei were isolated and allowed to incorporate radioactive dATP *in vitro*. After caesium chloride gradient centrifugation all the in-vitro DNA was part of the heavy DNA, indicating that the same pieces of DNA, which had begun replication *in vivo*, had elongated further *in vitro* (Schiebel & Schneck, 1974). This also showed that in-vitro initiation of replication does not occur and that DNA polymerase is already in the system.

A last comment on in-vitro DNA replication may be in place. It has been noted, that this system, employing homogenates, is sensitive to cycloheximide. This, and the fact that the outer nuclear membrane is studded with ribosomes, could suggest that these ribosomes are the sites of essential nuclear protein synthesis.

One further observation has recently been made which may be either an artifact or brings into question the organisation of the DNA in replicons. In the electron microscope, after a spreading of DNA molecules under almost denaturing conditions, many small 'bubbles', so-called microbubbles, have been detected which cover a stretch of DNA of about the length of an ordinary replicon. These bubbles are very small, down to 500 nucleotides, and if they were replicative structures, a 'replicon' would initiate replication almost everywhere at once, and the rate of replication which has been previously estimated at 0.2 μm/min would be about ten times too high (Hardman & Gillespie, 1980). However, the microbubbles, which have also been seen in rapidly dividing embryos, may indicate AT-rich or inverted repeats of DNA, which has already been replicated but whose daughter strands have not been ligated throughout the whole stretch of DNA. Even if shown to be an artifact, microbubbles should provide useful landmarks on the DNA.

RNA transcription

The elucidation of the mechanism of transcription in eukaryotic cells has produced some great surprises which could not have been expected from projection of the mechanism in prokaryotic cells. One is that all eukaryotes contain three distinct RNA polymerases, A, B and C, for rRNA, hnRNA (including pre-mRNA), and small (5 S and transfer) RNA, respectively. The other is that in addition to tailing and capping of mRNA precursors in an elaborate processing, some pieces get cut out of the RNA and the rest becomes joined thus creating mature mRNA. Consequently, eukaryotic genes can be mosaics of coding and non-coding pieces, named exons and introns.

As for the regulation of transcription, there may be a promotor sequence in front of the coding sequence, but it seems that chromatin configuration as well as DNA sequences within and far beyond the coding sequences play a major role, and that still-elusive 'factors' influence the polymerase enzyme, since an undefined cytosol extract must be added to allow for correct transcription (Weil *et al.*, 1979).

As the 'selective gene expression hypothesis' is still the only overall theory towards an understanding of development, we need to know all we can about *Physarum* RNA synthesis. Inhibitor studies have not been very meaningful, since actinomycin D acts only in almost saturated solution, and α-amanitin, specific for inhibition of RNA polymerase B, does not act at all *in vivo*, although both inhibitors are efficient and selective *in vitro*. In-vivo labelling with radioactive precursors, however, can be done very conveniently. In combination with in-vitro transcription in isolated nuclei and nucleoli, this has yielded some worthwhile results.

The total RNA of a plasmodium consists of about 82% rRNA, 1% 5 S rRNA, up to 10% tRNA, 0.1% 7–8 S RNA (small nuclear RNA) some circular RNA of unknown function, and about 1–2% of poly(A)-containing RNA, comprising the labile hnRNA in the nucleus and the cytoplasmic RNA, including mRNA. This RNA composition, reflecting the steady-state concentration of the various species, is quite different from the pattern of RNA synthesis where stable and labile RNA assume about equal proportions. A number of nucleases have been described which may act somehow after transcription, particularly on the labile RNA fractions.

Given the unique situation of distinct ribosomal genes in *Physarum*, the synthesis of rRNA is best known. If we take the electron microscope spread at face value, the primary transcript of rDNA is 4.2 μm long. This corresponds to 4.2×10^6 daltons of RNA and is quite close to the value for the precursor molecule of 4.1×10^6 daltons that has been isolated after in-vivo labelling. In the electron microscope this molecule has very little secondary structure, which is quite different from other pre-rRNA molecules, such as those in *Xenopus*. The small rRNA is directly cut from the precursor, it is a little bigger than the analogous species in mammalian ribosomes (19 S or 0.7×10^6 daltons or 2.1 kb). The large subunit seems to be generated via an intermediate size RNA of 1.9×10^6 daltons (34 S); when this molecule is fully developed, it is mid-way between bacteria and man in size (26 S, or 1.29×10^6 daltons or 3.9 kb). Upon heating a small molecule of 5.8 S (5000 daltons) is released. All this is similar to other cells, but a surprise comes from some r-loop mapping, where a piece of denatured rDNA is allowed to hybridize with mature rRNA. In the electron microscope it can be seen that the 19 S RNA forms a hybrid which allows localisation of the respective part of the gene on the rDNA molecule. However, the 26 S rRNA does not form a coherent hybrid: at two places, stretches of DNA find no partner, which results in loops of unpaired DNA of 1 kb and another one of 0.5 kb. Consequently, there must be two introns inside the structural gene for the 26 S rRNA, so far a unique situation (Fig. 18*b*). Of course this could be an artifact, if the rRNA is transcribed from a coherent gene and the electron microscopic picture happens to be from an exceptional gene. However, biochemical evidence has shown that at least 80% of the rDNA has these introns, and if the remaining ones do not take over all the work (which is unlikely from computing transcription rates *in vivo*), 26 S rRNA in *Physarum* arrives by a splicing mechanism.

Despite the electron microscopic and biochemical evidence, it is still possible that the nascent rRNA, or the structural gene for rRNA, is much longer and is being processed during transcription. Such a situation occurs in *Dictyostelium*. Also, it will be very important to identify the start sequence on the DNA, the putative promotor. Here, in-vitro studies have given new results. Although rDNA or isolated chromatin cannot initiate transcription *in vitro* with isolated RNA polymerase A, the endogeneous RNA polymerase enzymes are active in intact nuclei or nucleoli and in the mini-chromosomes mentioned above. While the latter produce transcripts of the correct DNA strand and the selective structural rDNA sequence, the pieces are very small. The former systems have been claimed to produce full-size transcripts (40×10^6 daltons) and achieve some processing. They even allow for some re-initiation *in vitro*. This has recently been utilised by incorporating a sulphur-containing nucleotide as the very first nucleotide. By affinity chromatography the newly started molecule could be isolated, and it hybridised to selective fragments of the rDNA. The site of initiation could be located on the map within about 3.5 kb away from the 19 S struc-tural sequence (Sun, Johnson & Allfrey, 1979). Just how far upstream the site of initiation is located is not yet clear, but seeing that this piece of DNA is being cloned a detailed DNA sequence will soon be known (Fig. 18*b*).

This still leaves a large portion of the central rDNA molecule for the untranscribed 'spacer', and with appropriate restriction enzymes the active and the inactive parts of the minichromosome can be cleanly separated into two test tubes. As expected, the active portion is more sensitive to DNase I and yields peak A after *Staphylococcus* nuclease digestion (Johnson *et al.*, 1979). However, this view has been challenged in a somewhat different experiment which claims that in comparison to mammalian chromatin the chromatin of *Physarum* is generally more sensitive to DNase I (Seebeck & Braun, 1980). This is not a trivial point, since a similar claim has been made for yeast, and we may learn something about the evolutionary position of *Physarum* besides all the special molecular biology of a gene that is rather boring in the eyes of some developmental biologists. However, the regula-tion of that gene in the life cycle is anything but dull, as will be shown below.

Transfer RNA (tRNA) is of course a mixture of many different species, each specific for one amino acid. Up to now 44 have been identified in *Physarum* – a small number considering that there are over 60 codons. Of particular interest here is that there are four codons for valine, yet only one valine tRNA. As far as is known, the activity of amino-acyl-synthetases, which could play a role in controlling translation, does not change during development. If the overall numbers are correct, we can expect about 25 genes for each tRNA in the genome (Melera, 1980).

The existence of mRNA has been shown only in 1979 by in-vitro translation, utilising a system derived from reticulocytes (Melera, Davide & Hession, 1979). This shows the problems involved in isolating undegraded RNA, free of the contaminating polysaccharide which is inhibitory in in-vitro experiments. The problem of preparing good polysomes has been another drawback in the assignment of mRNA sequences to the nucleus or the cytoplasm.

One major question – again with respect to the classification of a lower eukaryote, but also reflecting the degree of post-transcriptional controls – addresses the size of the primary transcript. Once again (and as the case of the rRNA has shown, with some justification), the electron microscope will give a direct answer (Fig. 19). Quite frequently Miller-type 'Christmas trees' are detected, and they all measure about 1–2 μm. This value translates into 2×10^6 daltons and is quite close to the value of 30 S (or 1.4×10^6 daltons) that has been determined in carefully extracted, labelled RNA by various groups of workers, either as pulse-labelled nuclear RNA or in vitro products of α-amanitin-sensitive material. While these estimates stress maximum size, a mean value (number average) has been estimated as 1550 nucleotides for the nuclear RNA. It seems therefore that hnRNA is not very long, though definitely larger than mRNA, and that the poly(A) is attached directly at transcription, as in all other systems. The mRNA has been estimated as 5–8 S or 1350 nucleotides. Nobody seems to have looked for the cap structure at the beginning of the mRNA, nor have any specific genes been analysed by cloned probes. With the great similarity of actin protein sequence between *Physarum* and *Dictyostelium*, somebody is probably probing the *Physarum* genome right now. Today no information on splicing of structural genes can be given, and all information to date is restricted to mixtures of hnRNA.

Aside from the size difference, there seem to be qualitative differences between hnRNA and mRNA. When nuclear or polysomal A^+RNA are hybridised with an excess of DNA, the latter takes a longer time to react. The hnRNA reacts faster, and about 30% hybridises before the mRNA fraction begins to react. These dot-curves can be interpreted in that hnRNA either contains separate transcripts of repeated DNA sequences, or that a repetitive DNA transcript is part of the hnRNA molecule and is cut out during processing (Fouquet & Sauer, 1975).

The kinetics of hnRNA processing, transport and its life-time in the cytoplasm are not well established. Rough estimates are 20–30 min for getting out of the nucleus at a loss of about two-thirds, and a half-life of 5–10 h later on, but more sensitive mRNA species may exist. The poly(A) stretches on RNA have been shown to consist of two size classes: one contains a rather long stretch (65 units) and seems to be on a stable ribonucleoprotein particle, the other is shorter (15 units) and unstable with a half-life of 4 h, which may be typical for the polysomal mRNA. It has been claimed by injection experiments that the ribonucleoprotein particles contain the putative mRNA.

The total amount of poly(A) RNA equals 1–2% of the total RNA. So far only one estimate is available for its qualitative composition in terms of sequence complexity. By determining the hybridisation kinetics of A^+RNA to cDNA (DNA which has been copied from that A^+RNA by reverse transcriptase), the following observations have been made. The main result is that *Physarum* RNA is quite complex. About one-half of the A^+RNA has a complexity of 3×10^7 nucleotides. If we take the estimated length of mRNA, this RNA mixture from *Physarum* (each molecule present in about 5 copies/nucleus) could code for 20 000 different proteins. The other half of the A^+RNA has an information content of a few hundred proteins, and each molecule is present about a few hundred times per nucleus (Baeckmann, 1980). It is surprising that no highly abundant stable A^+RNA species can be detected, in view of the fact that very intensive transcription can be seen in the electron microscope spreads, and that some proteins in the cytoplasm are very prevalent (actin up to 20% of the total soluble protein).

The high complexity of the RNA has been confirmed by an independent approach, hybridising A^+RNA to highly labelled single-copy DNA. This is quite similar to a previous hybridisation result with labelled RNA (Fouquet & Braun, 1974). It is currently not clear how much of this information, which corresponds to about 20% of the genome, is really utilised in the form of true mRNA on the polysomes.

In-vitro studies have been successfully done with intact nuclei, and the effects of α-amanitin and salt concentration have yielded some results on the template activity of endogeneous RNA polymerase B at various times of the life cycle. In another approach labelled α-amanitin has been used to titrate this enzyme (Pierron & Sauer, 1980a), and in a different way DNase I sensitivity has allowed an estimation of 15% euchromatin, a value not far removed from the RNA complexity data (Jalouzot *et al.*, 1980).

Finally, conclusive evidence for a different organisation of the fraction of the chromatin that is active in transcription comes from the electron micro-scopic spread preparation: putative active chromatin is not beaded but smooth (Scheer, Zentgraf & Sauer, unpublished).

The higher order organisation of chromatin in *Physarum* has been difficult to investigate. Sonication, and chelating agents like EDTA, have been required to disrupt the nuclei, but these treatments have disturbed the struc-tural integrity, even that of the nucleosomes. However, by destabilisation of

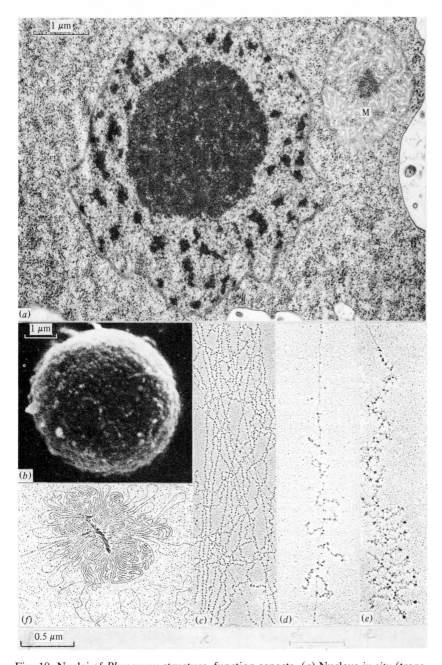

Fig. 19. Nuclei of *Physarum*: structure–function aspects. (*a*) Nucleus *in situ* (transmission electron micrograph). m, mitochondrion. (*b*) Nucleus after isolation (scanning electron micrograph). (*c*–*e*) Chromatin, after spreading (Miller-technique): (*c*) inactive, (*d*) transcribing putative mRNA, (*e*) transcribing putative rRNA. (*f*) Chromatin after DNase treatment. (Photos by G. Pierron, O. Schneider & R. Wolf, Würzburg.)

the nuclear envelope with lysolecithin the nucleus unfolds, and the nucleo-somal organisation is retained as well as endogeneous RNA polymerase activity (Schicker, Hildebrandt & Sauer, 1979).

This leads us to a brief discussion of RNA polymerases (see Seebeck & Braun, 1980, for review). Two enzymes, the nucleolar enzyme A and the α-amanitin-sensitive enzyme B, have been isolated in several laboratories and purified extensively. Enzyme C is suspected to be present at an early stage of purification, but no active enzyme has been isolated, nor is it detected in isolated nuclei.

At the latest count, RNA polymerase A with high specific activity of 650 U/mg is made up of five subunits, and RNA polymerase B of up to seven. As in other eukaryotic cells, there are two to three large subunits, and two of the small subunits may be common to both enzymes. Two forms of enzyme B may exist: both have one large subunit of 140 000 daltons, the presumptive binding site of α-amanitin; one form has a very large subunit of 210 000 daltons, the other a smaller one of 170 000 daltons. The second form may arise from the first one by limited proteolysis, probably not as an artifact but as a consequence (or even a cause?) of metabolic changes.

Three different 'factors' have been described that affect the activity of RNA polymerases *in vitro*. One of them is a protein which has been detected in homogenates of whole plasmodia, as well as in nuclei, but has not been purified extensively. It stimulates RNA polymerase (A or B) activity, provided a long double-stranded DNA molecule is used as template at low ionic strength. This 'elongation factor' might stabilise the transcription complex as it yields larger in-vitro transcripts.

Another factor has been found in the nucleolus. It is a molecule which sediments at 4 S and contains phosphate and carbohydrate (probably glycerol) in an unknown linkage. This material has a high affinity for free RNA polymerases and inhibits enzyme A more efficiently than enzyme B. Because it does not interfere with enzyme molecules that are bound to chromatin (i.e. engaged in transcription), it has been called 'initiation inhibitor'.

The third factor is a positive control element of the minichromosome. It is a protein of 70 000 daltons that binds very tightly (binding constant, 10^{-10} M) at the centre of the rDNA, but only in its phosphorylated form. If bound, it stimulates transcription of rRNA several-fold *in vitro*. The degree of phosphorylation of this protein is intricately regulated *in vivo*. Phosphoryl-ation is stimulated by polyamine and inhibited by cyclic AMP. How could the stimulation of rRNA synthesis by a protein such a long way away from the initiation site work? One interesting possibility has been suggested: the RNA polymerase molecules enter the minichromosome and line up in the spacer region of the rDNA in a bound, yet not engaged, state. The transition to the active state, i.e. proper initiation, could be regulated by conformational changes in the chromatin, caused by the phosphoprotein. Although this is speculation, we shall see that rRNA synthesis is controlled during the life cycle and, of course, rRNA limits the number of ribosomes and thereby the capacity for protein biosynthesis.

Protein synthesis

The protein contents of total plasmodia, amoebae or subcellular fractions have been quantitated in many cases. Subfractions of protein, after solubilisation by various treatments (acid extracts, increasing salt concentration, hydroxyapatite), have been displayed mostly on one-dimensional gels. The pattern of histone synthesis has been determined throughout the life cycle. The number of enzymes that have been investigated approaches 100, and some iso-enzyme patterns have also been analysed. In one case an antibody has been used to quantitate myosin. Labelling of nascent polypeptides has not been efficient with amino acid mixtures, but can be done with [^{35}S] methionine. Labelling with ^{125}I *in vitro* has been used to demonstrate residual protein on 'deproteinised' DNA.

An elegant method has been applied to distinguish whether an increase in enzyme activity has been due to activation or de-novo synthesis (Hütterman, 1973). After incubation with heavy amino acids and centrifugation of the homogenate on a density gradient, only freshly synthesised enzyme will be heavy.

One inhibitor of protein synthesis, cycloheximide, has been a very efficient drug to use in asking whether, and up to what moment, protein synthesis is required for a certain developmental process. Fortunately, a cycloheximide-resistant mutant is available to check whether or not the cycloheximide effect may have been an unspecific side effect of the drug.

As one further approach, many temperature-sensitive mutants have been characterised, and their effect at the restrictive temperature (38°C versus 26 °C) has been taken as the period in which the respective gene product has been required for its function.

In no case, however, has the whole story of a single polypeptide chain of *Physarum* been told from the transcription of its respective gene to its post-translational modification and eventual degradation.

After these general chapters on the subcellular composition and the molecular biology of *Physarum*, which can be viewed as necessary background, let us turn to the work on growth and differentiation of *Physarum* and stress developmental aspects. Before we move on, Fig. 19 provides a summary of some of the structural–functional aspects implicated above and elaborated below.

5

The growing plasmodium

Each microorganism or unicellular organism must multiply; and if it tried to do so without prior growth its progeny would become smaller and smaller until they disappeared. It seems very logical and economical if a cell divides just at the moment it has doubled its mass. Indeed, a critical size (or mass) has been measured in many cases (Mitchison, 1971, for review). The trouble is, how does the cell 'know' when it has grown enough? The classical theory is that a constant and a variable parameter relate to each other somehow, and cell division is triggered when a critical quotient is reached. This is the 'Kern-Plasma-Relation' of Hartmann, and the quotient of nucleus and the cyto-plasm has satisfied the prediction quite well (Prescott, 1976, for review). In the original version, volumes were related to each other; today the amount of DNA and total protein can be more accurately measured. This is not easy for a single cell, but it has been done and the respective quotient, protein/DNA, is quite constant among many individual cells in a population. Consequently, a random population can be analysed collectively. As a result, a growth curve can be constructed by either counting individuals over a long period of time, or measuring the weight or amount of protein, RNA or DNA. If adequate nutrients are provided, the growth is exponential, at least for a while. Semi-logarithmic graphs reveal a typical lag phase followed by logarithmic and stationary phases (Fig. 20). An individual in stationary phase will readily resume exponential growth, if put in a new culture dish. During the exponential phase the population is 'balanced', i.e. the rate of increase in cell number per unit time is constant, so is the quotient of protein/DNA or RNA/DNA. In a poor medium the slope of the curve will be slower than in a rich medium, because it takes longer to reach the critical size, but the population is still 'balanced' for each different growth rate. Whenever the cells arrive at stationary phase, they become 'unbalanced'.

Cells inside a multicellular organism are different in that they do not necessarily have to divide to sustain the organism. Many cell types stop divid-ing when they become differentiated; others, early in embryogenesis, turn into stem cells that can quickly respond to mitotic triggers, even if they do not obey the Kern-Plasma-Relation. Others grow bigger than predicted from the Kern-Plasma-Relation and fail to divide. Also, many cells die during ontogenesis and adult life, and are replaced in an ordered fashion.

66

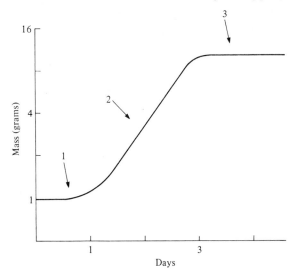

Fig. 20. The growth curve. A suspension of microplasmodia displays three typical phases after inoculation: 1, lag phase; 2, exponential phase, and 3, stationary phase. Microplasmodia in early exponential phase fuse well and yield macroplasmodia with good mitotic synchrony.

Therefore, the proliferation of cells within a metazoon must be carefully controlled, which means in most cases, that the potential to divide becomes restricted. Consequently, the growing process of a multicellular organism is far from being 'balanced', in contrast to a population of 'cycling' unicellular organisms. In fact, it has been questioned that higher cells cycle at all (Smith & Martin, 1974). At any rate, cells embark on a long lineage, and even if they divide, they do it in an individual way, for instance a line of blood cells or muscle cells.

However, animal cells can also be grown in tissue culture, where they behave as individuals. There are two types of culture: primary cultures and established lines. In the first case cells divide a few times, and then they stop growing, even if put back in fresh medium. In the second case, one cell of many is capable of sustained growth and while the rest die off its progeny populate the culture vessel and can be subcultured. In many cases the exceptional cell, starting a cell line (a clone), does not have a full set of chromosomes. As a rule, tissue culture cells, even if they proliferate well, are difficult to keep in a state of balanced growth. They stop growing when they touch each other, even if the nutrients are plentiful. This may be a control mechanism operating in the normal animal and is called 'contact inhibition'. It has been disturbed in some cancer cells which do not show this phenomenon.

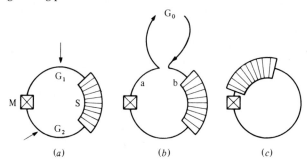

Fig. 21. Various formal mitotic cycles. (*a*) The typical cell cycle has four stations: mitosis M, G_1, S and G_2 phases, and transition points in G_1 and G_2 (arrows). (*b*) An extended G_1 phase can be formally subdivided into a G_{1a}, before the cell moves to the G_0 state, and G_{1b} after it re-enters the cycle. (*c*) Some mitotic cycles, like that of *Physarum*, lack G_1 phase.

It is an important observation that almost all cells stop multiplying temporarily or differentiate after mitosis but before the replication of their DNA. They are 'post-mitotic', or in a G_1 phase (or G_0 if G_1 lasts very long) in cell cycle terminology (Fig. 21). This argues for a control point early in the cell cycle, at which it is determined whether a cell will go ahead and divide, wait for some exogeneous trigger, or become differentiated. At that point the cell has not grown much and cannot know how the nutrient situation will be many hours later. This is in contrast to the arguments presented above where cells, such as protozoa, divide when they have doubled in mass and of course must have fully replicated their genomes.

Since there are good examples for both a general coordination of overall growth with cell multiplication and proliferation control early after mitosis, there may be two different principles regulating the mitotic cycle. Only those cells that can afford uncontrolled proliferation divide as frequently as possible. They override the control point early after their birth and consequently have a brief G_1 phase or none at all. They also have to divide, after passing their check-point late in G_2 phase, and display exponential growth and are balanced as long as nutrients are present.

In a series of elegant studies on fission yeast, involving mutants of the cell cycle (cdc, for cell division cycle), it has been shown that both control points can operate, but that the post-mitotic one is cryptic under optimal growth conditions (Mitchison, 1977*b*). On the other hand, genetic analysis of the budding yeast has clearly shown an early check-point in the cell cycle, called 'start'. At that point the cells become arrested, either to survive unfavourable growth conditions or to prepare for mating in response to the so-called α-factor (Hartwell, 1974). The various cell cycles are sketched formally in Fig. 21.

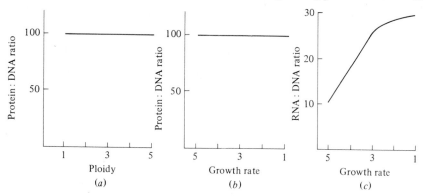

Fig. 22. Growth patterns in *Physarum*. The proportion of protein to DNA in plasmodia is stable at various ploidy levels (*a*) and at various growth rates (*b*), yet RNA increases over DNA as the growth rate declines (*c*).

All higher cells that have experienced a life history in a multicellular organism, must live up to their commitments before they make their DNA, i.e. they observe a check-point in G_1 phase, and may not ever divide again. Those cells do not grow exponentially, they are off-balance. Even if their proliferation control is disturbed, as in established cell lines and some cancer cells, they remain committed to their previous developmental history and express specific functions connected with their commitment (the so-called luxury molecules), over and above the general household programme.

There are few seemingly uncommitted cells in a higher organism, and consequently a pure cell cycle has only rarely been analysed independently from the cell differentiation that has occurred during development. Some exceptions may be the cell cycle of primordial germ cells, early embryonic stages (cleavage), and regeneration blastemas in plants and animals. In those systems a G_1 phase is short or absent. And, of course, there is the multinucleated plasmodium of *Physarum*, also with no G_1 phase.

Therefore, while the *Physarum* plasmodium may not be unique in displaying an undisturbed mitotic cycle, its natural synchrony and the ease with which it can be cultured in perfect balanced conditions make it well suited for analysis (Turnock, 1980, for review). What has been learned about the growth phase of *Physarum*?

A random culture of microplasmodia, when kept in a chemostat, keeps its balance over a five-fold change in growth rate, as its protein : DNA ratio remains constant at about 100 (Fig. 22*b*). The same constant value is found over a four-fold range in chromosome numbers (1n–4n, Fig. 22*a*). At a faster growth rate more protein must be made per unit time. As the capacity for protein biosynthesis depends on ribosomes, and as the number of amino acids that can be polymerised per ribosome and per second is rather constant at 100, the plasmodium needs more ribosomes at the faster growth rates. This is found by experiment, and the RNA : DNA ratio, although steady for one growth constant, increases from ten-fold to the maximum value of about 25-fold as the growth rate decreases (Fig. 22c). Knowing about the stable protein : DNA ratio and the increase in stable RNA, we can conclude that the amount of ribosomes not only controls protein synthesis during the cell cycle of *Physarum* but also sets the upper limit of how short it can get. These findings indicate that the synthesis of rRNA must play a major role in the control of the mitotic cycle.

On the other hand, it is difficult to imagine how protein mass can be accurately metered and how this can trigger mitosis. Therefore, one would like to identify a specific molecule, a 'division protein', that fluctuates during cell growth and triggers mitosis when it has reached a certain concentration (Zeuthen & Williams, 1969). This is a difficult task, as such a protein may occur as a minute fraction of the cytoplasm, and as long as such regulatory molecules are not known, it is not possible to select mutants that are defective in them. Indeed, very few mutants of the cell cycle parameters are available in *Physarum* or in any other type of higher cell. However, some cdc mutants in yeast clearly demonstrate, that cell growth can be uncoupled from cell division and DNA replication, as those mutants grow like wild types but do not divide, or divide too soon (Mitchison, 1977a). Although the respective gene products have not been identified, these mutants justify the search for specific mitotic 'factors'.

The cycling plasmodium

In *Physarum*, as in all other mitotic cycles, there are two decisive events. First, the genome has to double, and then the nucleus has to divide. The latter process can be accurately timed by phase-contrast microscopy, while incorporation of radioactive thymidine is a good indicator of the onset, at least, of DNA replication. This moment coincides with telophase which proves that there is no G_1 phase at all. Under optimum growth conditions, at 26 °C, it takes about eight hours for the following mitosis to set in, which defines the generation time. One major question has been to find out what must happen during that time period to ensure that mitosis can occur.

Let us first concentrate on the nuclei. At the time of birth, immediately after nuclear division, they are compact spheres of 4 μm in diameter, which quickly grow to 5 μm within 5 min, probably due to the decondensation of the chromosomes. During the following 50 min the nucleolus is reconstructed from granular remnants which have persisted throughout mitosis, and it takes on the central position as a sphere, covering about 5 μm^2 in a nucleus of 12 μm^2. S phase, according to radioactive-thymidine incorporation data, lasts about three hours. At that time a slight increase of nuclear and nucleolar volume occurs (Matsumoto & Funakoshi, 1978). The number of nuclear pores has been estimated at about 100 after birth, and this increases to 350 at 1 h and to 600–700 throughout the mitotic cycle.

Significant changes occur at 45 min before metaphase, when the nucleus abruptly grows from 20 to 30 μm^2, followed 20 min later by nucleolar growth from 6 to 10 μm^2. By that time the nucleolus has grown in mass too, since RNA and protein content rise from 0.5 pg and 3 pg to 1 pg and 4 pg, respectively. The enlarged nucleolus becomes kidney-shaped and moves towards the nuclear membrane. This stage can be well timed in routine experiments. It corresponds to about 20 min before the onset of anaphase.

At the centre of the nucleus a structure known as the microtubule organising centre (MTOC) has been detected, which may be derived from the nucleolus. Microtubules, 14–20 nm in diameter, radiate outwards from this structure, and these might be responsible for pushing the nucleolus to the periphery of the nucleus. The MTOC is rich in RNA, as judged from the fluorescence of acridine orange. During mitosis the MTOC divides in two, and these migrate to opposite poles of the nuclear periphery, forming the spindle poles of the intranuclear spindle (Laane & Haugli, 1974). In the electron microscope no definite MTOC structure can be seen, as microtubules originate at some distance from the spindle pole at the inner nuclear membrane and make contact with the chromosomes at the kinetochore, maybe one microtubule per chromosome. Microtubules that connect the poles have not been detected. The chromosomes swing around the nuclear centre for a few minutes (metakinesis), before they align in the metaphase plate. During that time the nuclei are frantically rotating in the protoplasm while the shuttle streaming goes on, although not as vigorously as in interphase (these are the 'unhappy' nuclei according to Zeuthen). Metaphase lasts 5 min and anaphase plus telophase about 3 min. During the latter period, the nuclear membrane is extensively deformed, and the nucleus becomes dumb-bell-shaped. While the nucleus is stretched (or pushed?) apart, temporal perforations may occur at the poles. The chromosomes move as a dark compact band towards each of the poles as the spindle fibres shorten, and a few minutes later two daughter nuclei are born.

The onset of anaphase is the most abrupt change that has been detected in a film analysis and serves as the most stringent marker of mitotic synchrony (Wolf, Wick & Sauer, 1979). It has been shown that neighbouring nuclei begin anaphase within 5 s, but the area under observation is limited. In isolated nuclei metaphase is a good indicator of synchrony. In a plasmodium of about 5 cm diameter containing 10^7 nuclei, over 98% are in metaphase within 5 min. In a larger plasmodium (10^8–10^9 nuclei) mitoses in different areas can vary by up to 20–40 min. The inoculum, which is the area from which a macroplasmodium grows out after it has been constructed by fusing microplasmodia, contains many nuclei that are out of phase. Further-

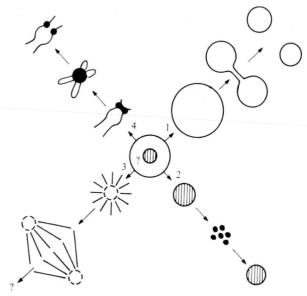

Fig. 23. Schematic dissection of some events usually associated with mitosis in *Physarum* plasmodia. 1. The nuclear envelope expands, as the nucleus swells and becomes constricted, leading to the dumb-bell shape before separation into two daughter nuclei. 2. The nucleolus swells, disaggregates and reaggregates. 3. An MTOC appears in the nucleus, divides and moves apart to construct the intranuclear spindle, and disappears after mitosis. 4. Replicated chromatin condenses into chromosomes, and chromatids decondense after telophase and start DNA replication.

more, if an established plasmodium is pressed down as a flat sheet in a culture chamber, such that nuclei cannot be moved around during the protoplasmic streaming, mitosis becomes asynchronous, although the plasmodium survives.

In summary, we can distinguish four discontinuous changes that are all associated with mitosis (Fig. 23): (1) an enlargement and division of the nucleus; (2) an enlargement, disintegration and reconstruction of the nucleolus; (3) appearance and division of an MTOC, together with assembly and disassembly of the mitotic spindle; and (4) chromatin condensation and decondensation.

The nuclear division in the plasmodium is not typical for higher cells, in that one observes an intranuclear spindle. In view of the fact that an amoeba of *Physarum* has a typical open mitosis with cytoplasmic centrioles (functioning as MTOC and spindle poles), it seems logical to explain the closed mitosis of the plasmodial organisation as a mechanism to avoid the mixing up of the chromosomes of adjacent nuclei after membrane breakdown.

However, as one of many examples to lend caution to our logical conclusions, the cellular slime mould *Dictyostelium* has a quite similar intranuclear mitosis to *Physarum*, but no plasmodial organisation. Yeast cells also have an intranuclear spindle, organised from the spindle pole plates at the nuclear membrane, yet in this case, no chromatin condensation occurs.

The main question remains: how is nuclear division regulated? Is there one trigger for the whole process, or are there separate 'factors' for regulating each of the morphological features of the mitosis, and how are they all orchestrated into the basic 'cell cycle'? Although many elegant experiments have been done, the picture is quite complex, as we shall now see.

Hunting for the mitotic trigger

The prettiest experiment is facilitated by the natural fusion of plasmodia (Fig. 24). In making a sandwich of two plasmodia, a joint 'cell' is formed with two kinds of nuclei, a heterokaryon. If the two plasmodia are in different phases of their autonomous cell cycles, a time-heterokaryon, is established, and an obvious question arises: do the mixed nuclei divide in synchrony, or does each kind of nucleus divide at the time when mitosis occurs in each donor plasmodium? In the many fusion experiments, since those of Rusch *et al.* (1966) (Holt, 1980, for review), the answer has been clear-cut: all nuclei divide in synchrony, and if the fused plasmodia are of equal size, the time of mitosis can be exactly predicted. If plasmodium A were to divide at 5 o'clock p.m. and plasmodium B at noon, the fused plasmodium (A/B) would divide at 2.30 p.m., which is just half the phase difference. This result shows that nuclei derived from plasmodium A will divide sooner, and those from B later, than those nuclei that have remained in the original plasmodia.

If a large piece of A were fused to B, synchronous mitosis would occur closer to 5 o'clock, by exactly the proportion predicted from the mass of each of the two plasmodia and their time interval. This rule of arithmetic averaging holds true, even if the two plasmodia are far apart in their cell cycle phases, for example early in S phase (about 1 h after mitosis) and late in G_2 phase (say 1 h before mitosis). In this case, at a cell cycle length of 10 h, synchronous mitosis will occur 4 h too late for the plasmodium in G_2 phase and 4 h too soon for the plasmodium in S phase.

These results demonstrate that throughout interphase some sort of intracellular communication can coordinate synchronous nuclear division. One can assume that some substance changes in concentration, and that mitosis is triggered when a threshold is reached. It is conceivable that the substance is used up at mitosis and has to be produced again in the next cycle. It seems straightforward to assume that the site of that material is in the cytoplasm, where it could become distributed evenly and interact with all the nuclei,

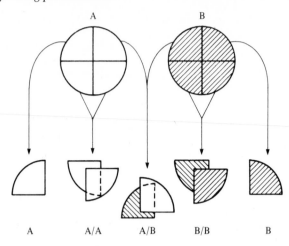

Fig. 24. The classical fusion experiment with *Physarum*. While nuclei in plasmodium A (or fused plasmodia A/A) divide in synchrony at one time, and nuclei in plasmodium B (or fused plasmodia B/B) divide also in synchrony at a different time, all nuclei in the fused plasmodium (A/B) divide in synchrony at the intermediate time (half-way between A and B).

assuring synchrony. However, since nuclei and cytoplasm are both mixed in fusion experiments and the trigger mechanisms are unknown, nuclei might interact in such a manner that they synchronise their clocks (by adjusting their DNA replication or their putative temporal RNA transcription mechanism).

Inhibitor experiments clearly indicate a decisive role for protein synthesis in triggering mitosis (Sudbery & Grant, 1976; Tyson, Garcia-Herdugo & Sachsenmaier, 1979). In many experiments cyclohexamide has been used. In one series the drug was added for a short period and then washed out again. The plasmodium recovered, and some delay of mitosis was observed when the drug was applied during S phase or early G_2 phase. However, a much longer delay occurred when protein synthesis was inhibited at 45 min before mitosis in late G_2 phase. This result has been interpreted in terms of protein that needs to be present at a high level in late G_2 phase, as a result of both production and degradation. While this observation may not convince everyone, it clearly argues against an inhibitor that needs to be synthesised in order to prevent mitosis, until it is diluted by general growth of the plasmodium to a low threshold value. In that case, cycloheximide would be expected to speed up mitosis instead of delaying it.

Other experiments with cycloheximide have shown that mitosis can still be blocked when the drug is added in prophase, just 10 min before metaphase, indicating the requirement of protein at a very late stage (Cummins &

Rusch, 1966). In the electron microscope, condensed chromosomes and radiating microtubules have been described, but no bipolar spindle. Thus the defect may be at the site of the MTOC.

Heat shock experiments, where the plasmodium is briefly held at an elevated temperature (37 °C instead of 26 °C, for 20 min) have led to somewhat similar results. A maximum sensitivity (expressed as mitotic delay), was observed at two hours before mitosis. Furthermore, high temperature applied at 15 min before mitosis for 10 min, blocked mitosis at metaphase or anaphase. Thereafter, the plasmodium returned to interphase (Brewer & Rusch, 1968). The disintegration and reaggregation of the nucleolus seems to occur at the normal time which leads to a ring of nucleolar material around the 'frozen' metaphase or a figure of eight around a 'frozen' anaphase structure (Wolf *et al.*, 1979). These results indicate a very late event in the cell cycle that marks the transition point beyond which the nuclei are committed to divide. This has been confirmed by clear-cut experiments asking how close to mitosis a nucleus can be forced back by fusion with a plasmodium in S phase. In these time-heterokaryons, the nuclei close to mitosis were prelabelled with [^{3}H]thymidine. The fused plasmodium was then incubated with bromodeoxyuridine to make heavy the DNA synthesised after mitosis. After a time correction, necessary for fusion and mixing to take place, it was shown that the nuclei from the advanced partner in the fused plasmodium were irreversibly committed at prophase, at 15–20 min before metaphase (Loidl, 1979). This transition point coincides with three further observations: (1) the high phosphorous content of histone H_1 which may cause condensation of chromatin into chromosomes (Fischer & Laemmli, 1980); (2) an intensive synthesis of $\alpha + \beta$ tubulin, known to accumulate in metaphase nuclei (Dove *et al.*, 1980); (3) the execution point of a temperature-sensitive mutant (MA 67) which blocks mitosis at 15 min before metaphase (at the restrictive temperature) and can be rescued by fusion with a wild-type plasmodium contributing only 5% in mass to the mixed plasmodium (Laffler *et al.*, 1979).

While these observations show a late phase of commitment, i.e. in prophase, the maximum delay after heat shock at two hours before mitosis indicates that other events in G_2 phase affect time of mitosis. Crude extracts from various phases of the cell cycle are most effective in accelerating mitosis when derived from nuclei two hours before mitosis; by about 10% of the cycle time if added in early G_2 phase (Chin, Friedrich & Bernstein, 1972). Similar results were obtained after partial enrichment of G_2-phase material. In this case, an antiserum against S-phase homogenate was used to precipitate S-phase components in a G_2-phase extract (Blessing & Lempp, 1978). Finally, a preparation of histone H_1 kinase (growth-associated histone kinase from a mammalian tissue) was added on top of a plasmodium, and a maximum acceleration (again up to 10% of cell cycle time) of mitosis occurred if the treatment was performed 3–4 h before the expected mitosis (Inglis *et al.*, 1976). It would seem very unlikely that an enzyme protein could survive uptake into a plasmodium and transfer to the nuclei. However, such a process has been clearly demonstrated for histones (Prior, Cantor, Johnson & Allfrey, 1980). These treatments may supply something which needs to be made prior to the late transition point of the mitotic cycle.

UV irradiation leads to a delay of the expected mitosis, if performed before nuclei are committed (the transition point corresponds to 50 min before metaphase). The earlier the irradiation the longer the delay, which can amount to 5 h. Overall growth is not affected, leading to an altered Kern-Plasma-Relation. Indeed, if crude extracts in late G_2 phase are added to such an irradiated plasmodium, the delay period is cut by one half (Matsumoto, 1977).

UV irradiation can also be employed to destroy a portion of the nuclei (Holt, 1980, for review). It is assumed that the UV light can only affect those nuclei which are close to the surface of the colourful plasmodium. These nuclei become pyknotic after the subsequent mitosis; they break up and disappear. The following cell cycle is shorter than in unirradiated controls. In UV-treated plasmodia the protein : DNA ratio is larger than in controls, and the shortened cycle conforms to the expected Kern-Plasma-Relation. The minimum generation time has been determined as 6 h by such experiments. The critical protein : DNA ratio could have been disturbed in two ways. Either there is too much protein, because a previous cycle was too long, or there is too little DNA, because some of the nuclei have died. It would be very important to know if a surplus of protein alone in one cycle can shorten the following generation time. If so, one could argue that the 'trigger' is not degraded at mitosis, but can be utilised once more in the following cycle. This, of course, poses a serious problem for the simple 'activator' model discussed above. Instead one would have to postulate an abrupt doubling of receptor sites for such an activator. These must then be 'titrated' with new activator molecules as the sites accumulate two-fold over a period that determines the generation time. One experiment is in favour of this interpretation. The drug fluorodeoxyuridine (FUDR), under appropriate conditions, can reversibly inhibit DNA synthesis. If this is done for either 3 h or 5 h, mitosis is delayed by exactly 3 h or 5 h, respectively. This indicates that S phase must be completed before mitosis can be triggered. As overall growth is not blocked, there is too much protein in the delayed plasmodium, and maybe too much activator as well since the subsequent cycle is shortened. This is consistent with the titration model just mentioned (Sachsenmaier, 1978). This experiment, taken together with the classic fusion experiments which revealed the 'averaging' phenomenon, predicts that the 'activator' accumulates throughout the generation time.

One further question has been investigated by yet another fusion experiment: is DNA replication required for the accumulation of the activator? In the control a plasmodium that had just finished S phase, say A, was fused with another plasmodium in G_2 phase, say B, and as expected synchronous mitosis occurred such that the nuclei from A were accelerated by one half of the time difference of the unfused plasmodia. In the experiment, plasmodium A was fused with a plasmodium *b* that had been kept in FUDR to block DNA synthesis until the time of fusion. On the face of it one would expect one of two possibilities: (1) if activator accumulates, while DNA synthesis is blocked, there should be a surplus of activator from plasmodium *b* and the nuclei from plasmodium A should divide sooner than in the control (2) if no activator is made when S phase is blocked, there should be no effect on the nuclei from plasmodium A, if fused with plasmodium *b*, and they should divide at the same time as in the control fusion experiment. The actual experimental result however is different: the nuclei of the fused plasmodium A/*b* divide in synchrony but much *later* than in the untreated (A/B), even later than mitoses in the original plasmodium A. Clearly, activator has not accumulated in the absence of S phase, and the delay of mitosis may indicate that the 'activator' left over from the previous cell cycle has been degraded during the blocked S phase. Consequently, the 'activator' seems to be a labile product of a gene, the expression of which is coupled to DNA replication. In other experiments a half-life for the activator of 1–2 h has been estimated (Tyson *et al.*, 1979).

Before considering the S phase, let us sum up the state of the hunt for the mitotic trigger. If one looks for *the* trigger, the data are very confusing (Fig. 25). If one just weighs the experimental observations and looks at the multiple events that are coordinated somehow for mitosis, there is no indication for a mitotic inhibitor; but also no evidence for a single activator. All we know is that there are several requirements for decisive events in S phase and G_2 phase and most prominently in prophase. Once this has been recognised, we may begin to ask the right questions.

Regulation of DNA synthesis

We have just shown: no S phase, no mitosis. Now we ask the converse question: is mitosis required for the occurrence of S phase? Another type of fusion experiment gives an answer (Rusch, 1970; Sauer, 1973; Holt, 1980, for reviews). In this case, local fusion is brought about by placing two flat plasmodia side by side and then cutting them apart only 15 min after fusion. Only a few thousand nuclei are exchanged, and their fate is followed after appropriate marking (radioactive labelling or nucleolar morphology) to distinguish between them. How do a few nuclei in G_2 phase respond to a large S phase plasmodium? Will they make DNA? The answer is no. Therefore, they seem to have to go through mitosis in order to begin DNA

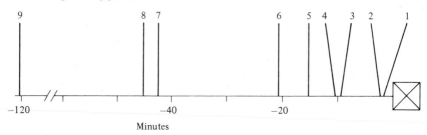

Fig. 25. Some events before mitosis. Various markers are listed, with respect to occurrence at estimated time before mitosis (×). 1, Slow down of shuttle streaming; 2, maximum pH value; 3, uncoupling of mitosis from S phase after heat shock; 4, blockade of mitosis after continuous cycloheximide treatment; 5, maximum phosphate content of histone H_1; 6, transition point for mitosis, according to fusion experiments monitoring DNA replication (BUDR-labelling); 7, transition point for mitosis, according to cycloheximide pulse experiments; 8, transition point after UV irradiation, delaying mitosis of the concurrent cycle; 9, maximal mitotic delay after heat shock.

replication. On the other hand, S-phase nuclei, put into a G_2-phase plasmodium, continue to make DNA. This shows that DNA synthesis is not repressed by an inhibitor in G_2 phase plasmodia, and a nucleus, once switched on to replicate DNA, becomes autonomous, at least for a while.

We have seen above that mitosis consists of several processes. Are they all needed to trigger S phase? The heat shock and cycloheximide treatments in prophase, both of which block a complete mitosis, nevertheless allow DNA synthesis. In the first case complete, although slow, DNA replication has been observed, leading to a doubling of the chromosomes (Wolf *et al.*, 1979). This procedure can be utilised to separate small from big polyploid nuclei in fused plasmodia. In the latter case, some 15% of the DNA is synthesised, which indicates that the onset of replication does not require protein synthesis while the completion of S phase does. Maybe the structural conformation of a condensed chromosome is enough to trigger S phase in *Physarum*.

In the intact plasmodium, DNA synthesis begins in all nuclei at the same time as shown by autoradiography. Incorporation of radioactive thymidine indicates that DNA synthesis begins at maximum intensity in telophase, reaches a plateau for 1.5 h, decreases slowly for a further 1.5 h and is low only after 3 h post-metaphase (Braun *et al.*, 1965). The low level of incorporation *in vivo* after S phase is due to nucleolar and mitochondrial DNA synthesis (Fig. 26*a*). The slope of the curve may indicate that DNA synthesis in all nuclei slows down at the end of S phase, or that some nuclei finish earlier than others, i.e. there is no synchrony at the end of S phase. These results must be viewed with caution: not only do the precursor pools

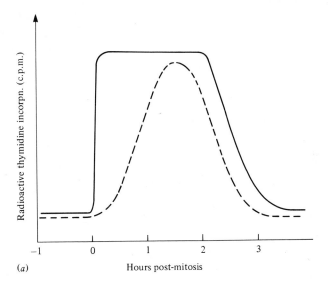

(a)   Hours post-mitosis

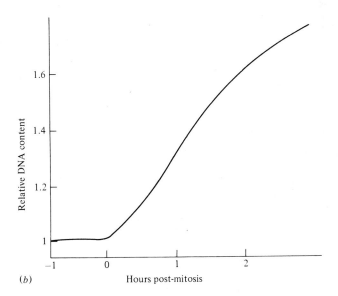

(b)   Hours post-mitosis

Fig. 26. Determination of S phase. (*a*) In-vivo thymidine incorporation (solid line), or in-vitro incorporation of thymidine monophosphate in isolated nuclei (dashed line); time in hours after mitosis (0). (*b*) Increase in DNA content after mitosis (0) computed by isotopic dilution of prelabelled DNA.

fluctuate (all dNTP pools expand just before mitosis), they seem to be sequestered in different compartments (nucleolar, nuclear, and cytoplasmic), and the most grave problem is posed by the fact that less than 25% of the labelled thymidine applied to the plasmodium appears in the DNA at all.

Therefore, *in vitro* incorporation serves as an important control. These experiments have shown that G_2-phase nuclei cannot replicate their DNA, nor can they accept exogenous DNA added as substrate for the DNA polymerases; S phase nuclei can do both. The results further confirm the abrupt initiation of replication, and the end of replication about three hours after mitosis. However, the capacity to make DNA reaches a maximum only after about one hour of S phase, in contrast to the situation *in vivo* (Fig. 26a).

A third approach has been to prelabel DNA and measure the decline of its specific activity, as fresh unlabelled DNA is made during the following S phases (Mohberg, 1974; Turnock, 1980). These isotope dilution experiments circumvent the problems arising with labelled precursor pools and can accurately determine the time course of DNA accumulation. The specific activity drops to about one half after the DNA has doubled. In one set of measurements a single plasmodium was followed for four consecutive cycles, and the prediction was found to be true, i.e. the plasmodium was really in balanced growth. These results confirm the absence of a G_1 phase and an increasing rate of DNA synthesis in early S phase as seen *in vitro* (Fig. 26b).

By 2.5 h 80% of the chromosomal DNA has replicated. An uncertainty which remains unsolved is that about 10% of the DNA at the end of S phase cannot be accounted for by this technique. Direct measurements of Feulgen-stained nuclei have claimed an increase in DNA content by some 10% during G_2 phase. This points to the possibility that S phase is much longer than has been suspected. The same argument comes from a consideration of the complex replication mechanism discussed above, which clearly indicates that ligation of full length DNA takes about 2 h. Consequently, even if we calculate DNA synthesis from incorporation data going on until 3 h after mitosis, completely replicated DNA molecules will only be present 5 h post-mitosis at the earliest. One extreme consequence of this is that there need not be a G_2 phase at all in the shortest possible cycle of *Physarum*. It may be significant that the shortest cycle time (after an increase in protein : DNA ratio or addition of some extracts) is 6 h, and, just as a reminder, in early embryos (such as those of *Drosophila* and sea urchin) neither a G_1 phase nor a G_2 phase seems to exist.

The autonomous synchronism of consecutive cell cycles in *Physarum* has allowed a uniquely elegant experiment, first done by Braun *et al.* (1965) (Holt, 1980, for review) to determine whether the DNA is replicated in a defined sequential order. A density shift of BUDR-labelled DNA is

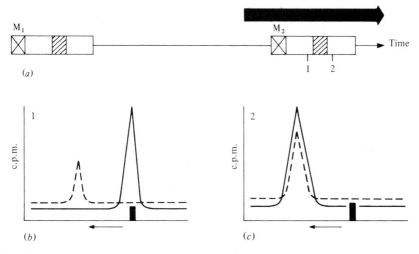

Fig. 27. Sequential DNA replication, as revealed by a density shift experiment. (a) A plasmodium was labelled for a brief period after mitosis, M_1, in S phase by [^{3}H]thymidine *in vivo* (hatched area). Before the following mitosis (M_2) [^{14}C]BUDR was added (solid arrow), which makes newly replicated DNA more dense. One sample of double-labelled DNA is analysed before (1) and one after (2) the time of prelabelling DNA with [^{3}H]thymidine pulse. (b,c) CsCl-gradient analysis of the labelled DNA: solid line, ^{3}H; dashed line, ^{14}C; arrow indicates increase of density, vertical bar marks normal density of unlabelled DNA. DNA sample 1 reveals non-coincidence of the two isotopes; sample 2 reveals coincidence of the two isotopes.

exploited in the following scheme (Fig. 27). During one S phase, DNA is labelled with [^{3}H]thymidine for a brief period; during the following S phase the DNA is labelled with a different isotope (^{14}C) and BUDR, either before, after, or at the same time as in the previous cycle. If DNA were replicated in the same temporal order in each S phase, only in the samples labelled at the same time interval would both isotopes be found in heavy DNA. These predictions have been met in a number of different laboratories: no doubt, DNA replication in plasmodia of *Physarum* follows a temporal sequence. In view of the DNA being organised in many replicons over the chromosome, it does not follow that the temporal order of the S phase reflects the physical or the geographical arrangement of replicons along the chromosome.

The DNA in the chromatin of a chromosome is about 2000 μm long and consists of about 100 replicons. We have learned that clusters of replicons (about four) are replicated together within about 1 h, and they are joined up with two other clusters, to make a big cluster of about 12 replicons each of which contains about 10% of the DNA or one 'bank of replicons'. If all clusters were to replicate in synchrony, the observed S phase would be too long; if one cluster were to follow replication only when the previous one had finished, the whole cycle would be too short to accommodate replication. Consequently, DNA replication may procede in isochrony among ten banks of replicons (Funderud *et al.*, 1978).

What keeps the banks in temporal order? It seems that protein synthesis is involved as cycloheximide specifically and drastically inhibits DNA replication, within minutes. It has been claimed that all three processes, initiation, elongation and ligation of DNA chains are affected (Schiebel, 1973; Funderud *et al.*, 1978).

In one set of complex measurements which have not been confirmed, cycloheximide was added to a set of plasmodia at ten consecutive points during S phase, and the accumulation of DNA seemed to follow as a stepwise pattern (Fig. 28). These results, taken together with the calculation of ten banks of replicons, might indicate that short-lived proteins must be made at 10-min intervals during S phase, which would drive the ordered replication of the genome. At this time it is not clear whether the cycloheximide effect reflects a general sensitivity of DNA replication towards protein synthesis capacity (a point raised by Bernstam, 1978), or whether it indicates that the synthesis of specific proteins is under the direction of some genes in the previous bank and is selectively required for the following bank of replicons to replicate.

One hint that the latter may be so comes from the following fusion experiment which has not been confirmed either. A large plasmodium in late S phase was fused with a small plasmodium in early S phase. Do the putative late proteins stimulate premature replication of late DNA in the early S-phase nuclei (Wille & Kauffman, 1975)? The answer is 'maybe', and the definite answer will come utilising cloned early and late DNA pieces to prove this claim.

In another experiment it has been shown that S-phase proteins do not accumulate when DNA synthesis is blocked. While DNA replication was interrupted by FUDR, without inhibiting protein synthesis, DNA replication was still sensitive to cyclohexamide after the release from FUDR, just as in untreated controls (Cummins & Rusch, 1966).

The highly ordered structure of DNA in chromatin will certainly be a factor in determining the sequence in which clusters of replicons can be replicated, and the unravelling of the chromatin in conjunction with the nuclear matrix and nuclear pores could be involved as well. The basic organisation of nucleosomes is not altered throughout the mitotic cycle. It seems to be basically the same in metaphase chromosomes as in S-phase and G_2-phase chromatin; actually, nucleosomes are assembled within 2 min after the replication fork has passed, yet small 'eyes' definitely contain less nucleosomes (Fig. 29).

One major question that can be uniquely addressed with the synchronous S phase of *Physarum*, is whether there is one specific DNA sequence where

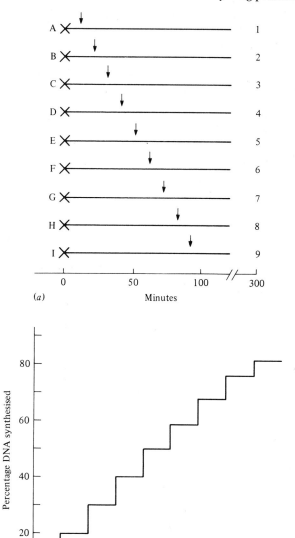

Fig. 28. A possible requirement for protein synthesis for sequential DNA replication in S phase. (*a*) A set of plasmodia (A–I) was treated with cycloheximide at regular intervals (arrows) after mitosis (×) (arrows), and the DNA synthesised in the presence of the drug was computed for the respective DNA preparations (1–9). (*b*) The proportion of DNA made in plasmodia A–I.

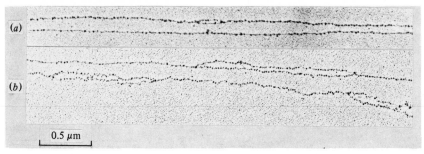

(a)

(b)

0.5 μm

Fig. 29. Replicating chromatin (Miller spread technique: G. Pierron & R Wolf, Würzburg). (a) Small 'eye' in replicating chromatin, seen in early S phase; (b) two 'eyes' and one 'fork'.

DNA replication can be initiated, something like the origin sequence in prokaryotes. If there is such a sequence, it has already been highly enriched by the following procedure (Beach *et al.*, 1980). At the very onset of S phase, the pieces of DNA which replicated first were density-labelled by incorporating BUDR. As the nascent DNA chains had not yet been ligated, they could be shaken out of the replicative structure by hand and isolated free of unreplicated DNA after a preparative centrifugation on caesium chloride density gradients (Fig. 30). Cloning of this material and DNA sequencing will answer whether or not there is an 'origin' in *Physarum*.

One last experiment on the initiation of S phase has shown that there may occur a 'master initiation' of all replicons in the genome of *Physarum* just at the start of S phase (Funderud *et al.*, 1978). In this approach, BUDR was incorporated at telophase. In this case the drug was then washed out, and after the whole S phase was completed, DNA was extracted and illuminated with black light (320 nm) to break the molecules which contained BUDR. One would have expected that the DNA of the first bank of replicons, approximately 10%, would be broken by this photolysis. Surprisingly, 100% of the mature DNA was broken. Consequently, the search for one master origin of replication may be successful.

These investigations, in summary, have shown that the S phase is a highly structured event (Fig. 31) and one may ask why. Further down we shall argue that this sequence of replication may actually contain and prescribe the order of gene expression in the cell cycle programme. Here, it should be added that early replicating DNA is heavier than the late DNA, contains less repetitive sequences, and it is more sensitive to DNase I digestion. Hence it may represent the euchromatin which needs to be transcribed during the growth of *Physarum* (Fig. 31).

One final comment on DNA replication refers to a covalent DNA modification. It has been determined that 5% of cytosine bases of *Physarum* DNA

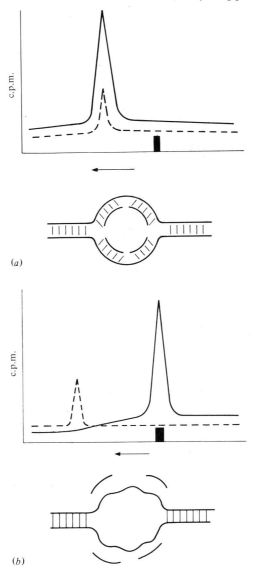

Fig. 30. Isolation of newly replicated DNA from early replicons. DNA was pre-labelled with [^{3}H]thymidine (solid line), and pulse labelled with [^{14}C]BUDR (dashed line) at the onset of S phase, and fractionated after CsCl gradient centrifugation (arrow). (*a*) Old and new DNA co-sediment at a position heavier than unlabelled DNA (bar). In the model (under the graph) new (i.e. heavy) DNA is attached to replicating DNA structure. (*b*) After vigorous shaking, new (i.e. heavy) and old DNA sediment as two separate fractions. In the model (under the graph), new DNA has been released from replicative DNA structure.

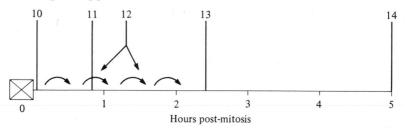

Fig. 31. Some events in S phase. Markers are listed against time after mitosis (×). 10, onset of S phase: (master-) initiation in telophase; 11, execution point of cell cycle mutant ATS 22; 12, sequential replication of banks of replicons; 13, end of DNA synthesis; 14, end of DNA ligation.

are methylated. Only 25% of the methyl groups are added at one S phase, and two more cell cycles elapse before full methylation is achieved. It also seems that methylation can occur on the parent strand, which is quite unusual judging by the work on other organisms (Holt, 1980, for review). In view of the possible role of methylation in differentiation, it will be important to find out more about DNA modification of the heterochromatin and other developmental stages of the life cycle.

As we have seen, there are two other kinds of DNA that need to be replicated in the mitotic cycle of *Physarum*, although on a per nucleolus or per mitochondrion basis it is not known whether all of the respective DNA molecules double. An exact control over the mitochondrial and ribosomal DNAs per plasmodium does exist.

It has been further shown that the amount of nucleolar DNA is kept at a stable proportion to the chromosomal DNA. About 175 genes per haploid genome have been computed for rDNA at various ploidy levels, growth rates and even in the haploid amoebae (Seebeck & Braun, 1980). However, two fusion experiments by the Gutteses have indicated that nucleolar DNA synthesis can be uncoupled from mitosis (Sauer, 1973, for review). In the first experiment a plasmodium was partially covered with paper, so that only the nuclei in the periphery underwent mitosis on time; those under the paper did not. In the second experiment a small, late G_2-phase plasmodium was allowed to fuse with a large, S-phase plasmodium, where the nuclei of the former were prevented from going through their mitosis. In both cases the nucleolar DNA synthesis continued, probably leading to an amplification of the ribosomal genes.

Mitochondrial DNA synthesis, as measured by radioactivity incorporation and autoradiography in the electron microscope, can be detected throughout the mitotic cycle. This was initially taken to be evidence for random replication. However, a careful study of shape, size and and distribution of mitochondria, and of labelled DNA over the nucleoids, has demonstrated a good, though not complete, synchrony of a mitochondrial division cycle (Goodman, 1980, for review). Mitochondria divide in the first half of the nuclear S phase, and after their G_1 phase of 3 h, replication of the mitochondrial DNA takes place during the G_2 phase of the nuclear cycle, even in the presence of cycloheximide. Mitochondria remain in their G_2 phase while the nuclei divide.

Clearly, the three kinds of DNA in a common plasmodium are regulated in different ways. We shall now consider what has been learned about the other functions of chromatin, the transfer of information from the nucleus to the cytoplasm, or transcription of RNA.

Regulation of RNA synthesis

Although a turnover of nuclei, including their DNA, occurs in the development of *Physarum*, these are much more stable than RNA; however, ribosomal RNA is quite stable. Therefore, the isotope dilution technique has been employed (Turnock, 1980). Indeed, the specific activity of mature 26 S + 19 S RNA falls exactly by one-half and no further during one full cycle, indicating that at least mature rRNA is stable and the plasmodium in balance.

When the rate of rRNA accumulation is calculated from the loss of specific activity at regular time intervals in the cell cycle, two observations are made (Fig. 32a): first, there is no rRNA increase at mitosis; second, the rate of accumulation is not linear but increases at the end of G_2 phase to five times the value measured in early S phase. Quite similar results have been obtained for tRNA, indicating that these two classes of RNA, both of which are involved in limiting the capacity for protein synthesis, may be co-ordinately controlled, although they are synthesised from chromosomal or extra-chromosomal genes by two different RNA polymerases, respectively.

However, we have described earlier that rRNA in *Physarum* is processed by a complex pathway including splicing. Therefore, the doubling in mass of mature rRNA as well as its increasing rate of accumulation may not be determined by the rate of transcription alone. This seems to be true, as the level of RNA polymerase A determined in isolated nuclei or nucleoli does not change during the cell cycle, and this does not support an exponential synthesis rate at the end of the cycle (Davies & Walker, 1978). In addition, a low level of RNA polymerase A is detected at mitosis, and an increase to the steady level has occurred by the end of nucleolar reconstruction, i.e. already in early S phase. Even in metaphase, at the lowest level of polymerase A activity, the enzyme has been shown to be present, though inactive. How this enzyme is inactivated at prophase and reactivated during nucleolar reconstruction *in vivo* is not known. It may be speculated that the massive influx of actin, which has a high affinity for RNA polymerase A (Smith &

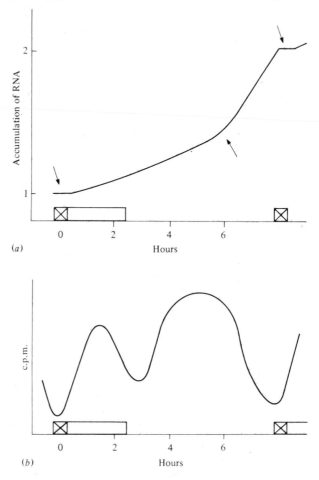

Fig. 32. RNA synthesis during the mitotic cycle. (*a*) Accumulation of stable RNA, as deduced from isotope dilution from prelabelled RNA. The arrows mark a halt in RNA synthesis at mitosis (×), and an increase in the rate of synthesis towards the end of the mitotic cycle. (*b*) Incorporation of radioactively labelled precursor (uridine) by pulse labelling reveals a biphasic pattern of RNA synthesis.

Braun, 1978), into the nucleus in late G_2 phase, impairs transcription, although other functions for actin in mitosis can be envisaged.

In-vivo RNA synthesis, conveniently analysed by radioactive uridine incorporation (Fig. 32*b*), shows a pattern quite distinct from that of rRNA or tRNA accumulation. It is biphasic, with a very low minimum at mitosis and a second minimum in early G_2 phase at about one-third of the two peaks of incorporation: one in S phase and one in G_2 phase (Rusch, 1970; Melera,

1980; Seebeck & Braun, 1980, for reviews). These results have been obtained after 10-min pulse labelling, when about 70% of the RNA is still in the nucleus. These fluctuations are probably not due to variable pools of precursors, since the same synthesis pattern has been confirmed *in vitro* with isolated nuclei. With restriction fragments as cloned pieces of rDNA at hand, the true pattern of rRNA synthesis, as opposed to its accumulation, will soon be known.

The composition of the pulse-labelled, labile RNA cannot be accurately given. Size differences in the pulse-labelled RNA have not been detected. Some indirect arguments, however, support the hypothesis that the RNA made in early S phase (from telophase to 45 min after metaphase) is rich in hnRNA. Its base composition is more DNA-like, whereas G_2-phase RNA contains more $G + C$ (i.e. it is more like rRNA). S-phase RNA hybridises with chromosomal DNA up to 40%, a value about three times that obtained with G_2-phase RNA. The proportion of poly(A)$^+$RNA in early S phase is several times higher than in G_2 phase, and it contains less RNA transcripts from repetitive DNA sequences.

Further evidence comes from in-vitro analyses of nuclear RNA polymerases. Using α-amanitin to discriminate RNA polymerase B, it has been shown that the enzyme is only active in S phase (Pierron & Sauer, 1980*a*). However, RNA polymerase is also present in G_2 phase, yet not active. This is concluded from the effect of high salt concentration on this RNA polymerase activity. Whilst in S phase this enzyme cannot be stimulated and seems fully active, in G_2 phase its activity can be fully restored by high salt concentration to the S-phase level (Grant, 1972). More direct evidence has come from titration with radioactive α-amanitin. Throughout the whole cell cycle the level of the enzyme has been found to be completely stable. However, it cannot be detected in metaphase nuclei (Pierron & Sauer, 1980*b*). This enzyme may belong to the so-called shuttle proteins, which leave the condensed chromosomes and return selectively at decondensation of the chromosomes after mitosis.

If the transcription of genes plays any role in cell cycle control, it would be expected to be realigned once every cycle. Metaphase may coincide with the moment of resetting the transcription programme, by eliminating all RNA polymerase B molecules which have been engaged in transcribing in the previous cycle. Fresh enzyme molecules become attached straight away, which for some reason are more active in S phase than in G_2 phase. As mentioned above, the enzyme becomes partially inactivated in G_2 phase. In addition, the 'elongation factor' (that protein fraction which stimulates RNA polymerases *in vitro*) is found at an elevated level in early S phase, adding a putative positive control element to transcription by RNA polymerase B. Furthermore, the S-phase chromatin is more receptive to homologous RNA polymerase B (Schicker *et al.*, 1979). This was observed when 'chromatin', prepared by the lysolecithin method, was found to accept isolated enzyme when it was derived from S phase, but not from metaphase or from G_2 phase; a second brief period of sensitive template conformation was found at the transition of G_2 phase to prophase.

The question can be asked: is there more than a temporal correlation between S phase and RNA synthesis? Looked at one way, the dependence of DNA synthesis on RNA synthesis is quite clear, though the amount of RNA involved is small and its synthesis is not understood: chromosomal DNA can only be replicated after primer-RNA has been provided. Looked at the other way, in *Physarum* the dependence of RNA transcription by RNA polymerase B on ongoing DNA replication is very obvious (Sauer, 1978, for review; Pierron & Sauer, 1980*a,b*), at least in quantitative terms. If DNA replication in early S phase is inhibited by FUDR or hydroxyurea, the incorporation of precursor into total RNA is significantly inhibited – by more than 70% in the case of poly(A)$^+$RNA. No significant effect of the drugs is seen in G_2 phase. As the drugs could still upset RNA precursor pools *in vivo*, RNA polymerase levels were determined in isolated nuclei, and were found to be completely (up to 100%) inactivated (Fig. 33*a*). Now the question arises as to whether RNA polymerase B is absent from chromatin blocked in S phase, or whether chromatin configuration needs to be 'open' to allow concomitant transcription. By titration with radioactive α-amanitin the normal amount of RNA polymerase B, just as in untreated controls, has been established (Pierron & Sauer, 1980*a*). Two indirect observations point to a decisive role of chromatin structure in template activity of RNA polymerase: one is that chromatin prepared by the lysolecithin method, which can be transcribed by exogeneous RNA polymerase in S phase, is not active if derived from a blocked S phase; and the other, that DNA in early replicating chromatin is selectively DNase I-sensitive only when replication is intact. Finally, and after what we have discussed about DNA replication not surprisingly, cyclohexamide treatment also affects RNA transcription selectively in S phase.

All the evidence is consistent with the hypothesis of replication–transcription coupling (Sauer, 1978). It is not clear however, whether the DNA in early S phase, presumably euchromatin, is really transcribed as it is replicated, thus determining a definite sequence of gene products. Some initial chromatin spreads have been observed which contain RNA fibrils within replicating 'eyes' (Fig. 33*b*). Further work, hopefully with cloned DNA segments, is necessary to prove whether replication–transcription coupling is more than a temporal coordination.

At this time, it cannot be said whether the intensive transcriptional activity in the first half of S phase is an essential event in the mitotic cycle, or just a sign of open chromatin, whose transcription cannot be stopped until G_2 phase has arrived (see Fig. 34). However, it is clear that gene dosage

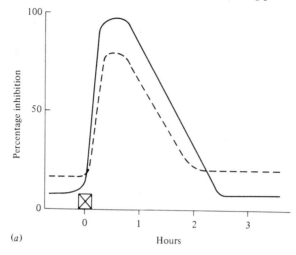

(a)

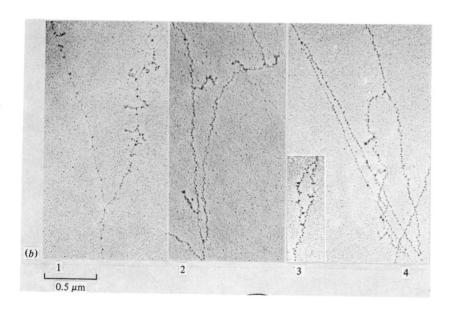

Fig. 33. Some evidence for replication–transcription coupling. (*a*) Inhibition of RNA synthesis (poly(A)$^+$RNA) *in vivo* (dashed line) or nuclear RNA polymerase B-activity (solid line) after halting DNA replication. (*b*) Four replicative chromatin structures, with putative RNA transcripts (Miller spread technique; G. Pierron & R. Wolf, Würzburg).

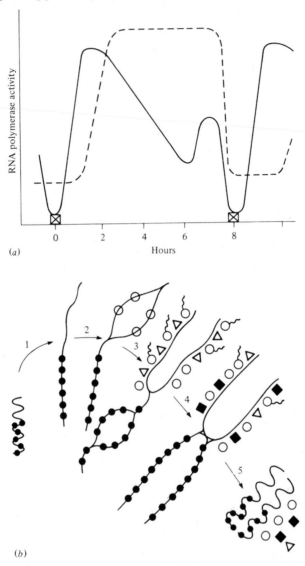

(a)

(b)

Fig. 34. Tentative activity patterns of RNA transcription during the nuclear cycle. (a) RNA polymerase A (dashed line), RNA polymerase B (solid line) between mitoses (×). (b) A sketch indicating: 1, decondensation of the chromosome after mitosis, of smooth and beaded (closed circles) chromatin; 2, early replication and some transcription by newly attached RNA polymerase B; 3, late replication and more RNA transcription on replicated chromatin through putative stimulatory factor (triangle); 4, reduced transcriptional activity, through a putative inhibitory factor (square); 5, condensation of chromatin into chromosomes (and release of RNA polymerase B and putative factors).

. cannot generally govern the rate of transcription in *Physarum*, since in G_2 phase, when the genes have doubled, hnRNA synthesis is clearly reduced. It is also unknown whether there are qualitative changes in the composition of the stable fraction of the poly(A)$^+$RNA, which amounts to about 1% of the total RNA throughout the cycle. As judged from an estimate of the sequence complexity and DNA/RNA hybridisation with single-copy DNA, there are no major changes (Baeckmann, 1980), and in no other cell cycle (beyond yeast) have distinct qualitative changes, indicative of selective transcription of mRNA, been described. The pattern of polypeptides after in-vitro translation of mRNA (contained in the total poly(A)$^+$RNA preparation) has not revealed obvious changes after one-dimensional gel electrophoresis (Baeckmann, 1980). This last observation raises the question of whether a more detailed analysis of protein synthesis can give new clues to the understanding of the mitotic cycle.

Regulation of protein synthesis

While the use of actinomycin D, an inefficient inhibitor of the synthesis of RNA classes, has not given a clear picture of the involvement of selective gene transcription in *Physarum*, experiments with cycloheximide have clearly shown that the progression of the cell cycle requires protein synthesis, i.e. the translation of mRNA. It is not clear whether the general interruption of the whole protein synthesis apparatus or the lack of specific polypeptides is the decisive factor in stopping the cycle. This is a general problem which has not been solved in either bacteria or yeast, despite detailed knowledge of their genetics (Halvorson, 1977). Therefore, it is necessary at first to study overall translation.

In *Physarum* total protein doubles between mitoses, and the accumulation of protein occurs linearly. However, an estimation of synthesis rates has revealed that protein synthesis in S phase is low and increases exponentially during G_2 phase. This increase at the end of the mitotic cycle correlates with the rRNA pattern. It may well be that overall protein synthesis, still the best parameter in describing nucleo-cytoplasmic ratios (Kern-Plasma-Relation), does depend on the availability of ribosomes, which would make rRNA synthesis a major control factor of mitosis (Turnock, 1980).

The amount or synthesis of individual polypeptides, as revealed by stained or labelled bands in electrophoresis, has not shown distinct changes in various kinds of protein extract derived from whole plasmodia, nuclei or nucleoli throughout the mitotic cycle (Walker *et al.*, 1980), with three exceptions. (i) Actin accumulates in the nucleus in late G_2 phase (Jockusch, Brown & Rusch, 1970); (ii) tubulin synthesis increases 50-fold abruptly and briefly in prophase, just before the spindle is formed (Dove *et al.*, 1980); (iii) histones, collectively, are present in about an equal amount to DNA throughout the mitotic cycle, indicating that they are made only while DNA is replicated. This latter point has been confirmed by radioactive incorporation studies and may be one example of replication–transcription coupling (Mohberg, 1974). A tight coupling of histone to DNA synthesis is not apparent, as histones are also made when DNA synthesis is blocked by FUDR, and furthermore, a pool of free histones is available. This has been concluded from isotope dilution experiments: while the specific activity of DNA decreases exactly by one-half to 0.5, the value for histone reaches only 0.7 after the next S phase, indicating that labelled histone from the previous cycle gets into the replicating chromatin in the following cycle (Mohberg, 1974).

Most enzymes that have been analysed in *Physarum*, follow a continuous pattern, as they are presumably made throughout the mitotic cycle. This is true for the polymerases of DNA and RNA which at first sight might be considered regulatory factors (Sauer, 1978, for review). However, as discussed above, the fraction of the enzyme that is bound to the nuclei fluctuates considerably. Quite similar results have been obtained in *Escherichia coli* (Lutkenhaus *et al.*, 1979) and yeast (Elliott & McLaughlin, 1978), where even after short pulses over 500 protein spots were present at all times of the cell cycle after analysis on two-dimensional gels. Any exception to this rule, then, is of interest as it provides a 'landmark' in the cell cycle and becomes a potential candidate for a regulatory factor controlling mitosis.

Not every change in enzyme activity, however, is a cell cycle event, and most cases described in other systems which have to be synchronised for all cycle studies (Mitchison, 1971) turned out to be a consequence of imbalanced growth (Mitchison, 1977*a*). Also, with the great variation of enzyme activity, due to co-factors or post-translational modifications and the turnover of enzyme protein, an abrupt change in enzyme activity does not have to reflect directly transcriptional changes.

There are about ten examples of enzyme fluctuations in the mitotic cycle of *Physarum*. Among those are peak enzymes whose activity goes up and down again, such as thymidine kinase, deoxyadenosine kinase, NAD-pyrophosphorylase (in S phase), guanylate cyclase (shows one peak in S and another one in G_2 phase), adenylate cyclase, histone kinase (shows peaks in G_2 phase), histone phosphatase (shows two peaks, one in G_2 and one in prophase). There are some step-enzymes which abruptly double in activity in G_2 phase, such as RNase, glutamate dehydrogenase, polyATP-ribose polymerase. Ornithine decarboxylase has one step which occurs in early S phase (Mitchell & Rusch, 1973). This is the first enzyme in the pathway of the synthesis of polyamines (e.g. spermidine), substances that have been associated with growth control; it may also be another case of replication–transcriptional coupling.

Two peak enzymes deserve further comment. In view of the high phosphate content of histone H_1, and the acceleration of mitosis after application of a mammalian histone kinase preparation, and the hypothesis that chromosome condensation is an initial step in controlling mitosis, the endogeneous histone kinase could be causally involved in an essential regulatory process (Matthews & Bradbury, 1978). This enzyme was found in the nucleus, and its activity increases 15-fold, beginning after S phase and reaching maximum value at 2 h before mitosis. Further work on this enzyme has resolved two activities, one at 2 h and the other at 1 h before mitosis. This indicates that the high phosphate content in prophase does not correlate with the enzyme peaks, and it seems to be rather a balance of phosphorylation in and dephosphorylation by histone phosphatases which peak late in G_2 phase. Furthermore, by growing a plasmodium in heavy amino acids from mitosis to G_2 phase (over a period of 6.5 h) and separating the newly made heavy kinase protein by isopycnic centrifugation, only a doubling of the enzyme amount was found, which makes histone kinase a continuously synthesised protein and its activity another case of enzyme activation, and definitely not an example of selective gene expression in the cell cycle (Mitchelson *et al.*, 1978).

The other enzyme that has been thoroughly analysed in thymidine kinase. Its activity increases sharply just before mitosis, and its peak parallels S phase (Sachsenmaier & Ives, 1965). Cycloheximide inhibits the increase immediately, and density labelling has shown that the enzyme protein doubles within about 3.5 h. The inhibition of the enzyme activity increase by actinomycin C sets the time of transcription of the thymidine kinase gene at about 1.5 h before mitosis. This time of putative gene activation is close to the transition period deduced from the various perturbation experiments (see above, Fig. 25), and thus thymidine kinase has become the first available marker for 'commitment' towards mitosis. This tight correlation can be further seen from the coordinated delay of enzyme increase as well as mitosis at elevated temperature (32 °C) and from the two steps of thymidine kinase activity, when a plasmodium is constructed, in which two sets of nuclei divide after an interval which equals exactly that of the enzyme activities (Wright & Tollon, 1979*a,b*).

Whether the enzyme is functionally involved in the procession of S phase, is a different story. Thymidine triphosphate (TTP) as a precursor for DNA synthesis is mainly derived from nucleotides other than thymidine, and thymidylate synthetase, the essential enzyme to make TTP, is a continuous enzyme. However, thymidine could be channelled into a sequestered precursor pool, which might control ordered replication. If so, thymidine kinase could achieve more than just salvaging thymidine in intermediary metabolism. Careful analysis has revealed three thymidine kinase activities with different iso-electric points; one is present at all times at a low level, another only in mitosis, and a third only in S phase. Further analysis has shown that the enzyme peak in mitosis is probably due to de-novo synthesis, and it is then converted into the form which is active in S phase, probably by a dephosphorylation (Gröbner & Sachsenmaier, 1976). Consequently, the tight coupling of thymidine kinase synthesis and mitosis can be traced back to gene expression. The possible function in S phase is controlled by a post-translational modification. As mutants deficient in thymidine kinase are available in *Physarum* (Lunn, Cooke & Haugli, 1977), which grow normally and have an undisturbed cell cycle, we can safely conclude that fluctuations in thymidine kinase (TK) during S phase are irrelevant for the continuation of DNA replication. But we can learn something else (aside from the proof that the three forms of enzyme come from one structural gene) from these TK mutants which can have less than 2% of thymidine kinase activity as compared to the wild type. Namely, even if a specific protein fluctuates 50-fold, we cannot be certain that it is really required to keep the cycle going. One last observation on the TK mutants stresses once more that the respective gene is in an interesting portion of the chromatin with respect to the 'commitment': the residual enzyme activity still increases at mitosis. It will be most rewarding to learn if the ts mutant (MA 68) with an execution point at 15 min before mitosis belongs to the same area in the chromatin (Laffler *et al.*, 1979).

Of course, only a combination of genetics and biochemistry will lead to a better understanding of the cell cycle controls. Genetic techniques have gradually been worked out (Haugli, Cooke & Sudbery, 1980). The TK⁻ mutants have been a by-product of a still unsuccessful hunt for DNA⁻ mutants; they are BUDR-resistant (the drug being used in the selection regimen). Searching for cell-cycle mutants in *Physarum* among mutants that cannot grow has been futile. Obviously, growth and the cell cycle are different developmental situations with two sets of genes in action. Hence, true cell-cycle mutants require a tedious screening of many young mutant plasmodia for DNA-, RNA-, or protein synthesis (Sudbery, Haugli & Haugli, 1978; Del Castillo, Oustrin & Wright, 1978), the mitotic index (Wheals, Grant & Jockusch, 1976), the nuclear reconstruction index (Gingold *et al.*, 1976), and nuclear morphology (Burland, 1978; Burland & Dee, 1980).

As an optimistic view, one can anticipate more meaningful facts on the participation of gene expression controls in the cell cycle of *Physarum*. For now we can merely sketch some discontinuous events concerning the cell cycle (Fig. 35).

Some small molecules

Although we have mainly been concerned with the macromolecules in the mitotic cycle, since they establish at the same time growth of the plas-

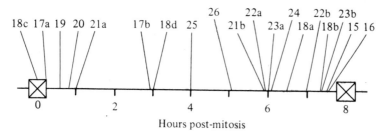

Fig. 35. More discontinuous events of the mitotic cycle. 15, Burst of α and β tubulin synthesis; 16, execution point of mutant MA 68; 17a, histone synthesis, switched on; 17b, histone synthesis, switched off; 18a, thymidine kinase (TK) transcription of structural gene; 18b, translation; 18c,d, post-translational modifications; 19, ornithine decarboxylase (ODC) synthesis; 20, NAD-pyrophosphorylase; 21a,b, cyclic GMP peaks; 22a,b, histone kinase (HK) activity peaks; 23a,b, histone phosphorylase peaks; 24, cyclic AMP increase; 25, RNase-step; 26, glutamate dehydrogenase step.

modium, small molecules may be important if they show large fluctuations.

The internal pH value cycles in phase with the mitotic cycle. It increases continually during most of the generation time; its highest value coincides with mitosis (pH 6.5), and it drops in early S phase (pH 6) (Gerson, 1978). The rate of oxygen uptake is not constant; it shows an increase in S phase, another in mid G_2 phase, it is steady at late G_2 phase plus mitosis, and in early G_2 phase (Forde & Sachsenmaier, 1979). The pools of amino acids and RNA precursors do not vary considerably, while the dNTP pools expand about two-fold at mitosis (Turnock, 1980). A number of elements have been determined in the plasmodium and in nuclei (Jeter *et al.*, 1979). While chloride (Cl^-) does not fluctuate, most elements show low concentrations in S phase and higher ones at mitosis. No major differences have been observed between cytoplasm and nuclei, with the exception of metaphase nuclei which contain more phosphorus (P), magnesium (Mg), and sulphur (S).

It is of course not clear whether the fluctuations are related to mitosis or other events in the plasmodium, and as we shall see below it is quite important to ascertain what fraction of a particular element, say calcium (Ca), is sequestered, free or bound in the motile apparatus.

Before we come to other aspects of the plasmodium, a brief look at models of mitotic controls may be necessary, even if they may have been more misleading than helpful so far.

Mitotic models

In the previous sections we have discussed some of the observations made and experiments done on a growing plasmodium, work that can be sum-

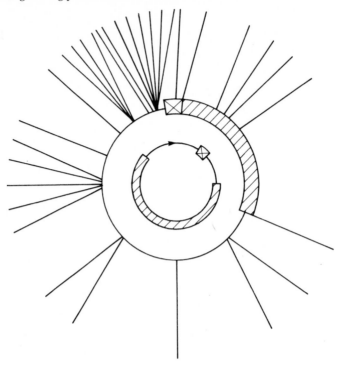

Fig. 36. Summarising events of the mitotic cycle of *Physarum* (cf. Fig. 21). The outer circle represents the nuclear division cycle (mitosis, S phase and G_2 phase), the inner circle the mitochondrial division cycle (division, G_1 phase, S phase and G_2 phase); the lines indicate the events detailed in Figs. 25, 31 and 35.

marised as cell-cycle studies. If we draw in a circle all the markers and discontinuous events (Fig. 36) that we can detect, we obtain an impressive picture of the mitotic cycle; it is also a very complex picture with events clustered in S, in G_2, and in the G_2/M transition-period (commitment) and at mitosis. Although this sketch looks like a sun, it may not be quite that enlightening! Once more we repeat that several oscillations occur during the cycle, among them those of the chromosomes, the nucleolus and the mitotic apparatus in the nucleus and, in a different phase, the mitochondrial cycle. So it may not be surprising that the picture is complex. But at this point it seems to be more that it does not make sense. This is, where models come in handy, because they do not have to reflect all the facts; they start out simple and can be adjusted as they are tested – models should be testable, though.

The model which has been published to account for mitosis in *Physarum*, has been called the 'hour glass model' (Sachsenmaier *et al.*, 1972; Sachsen-

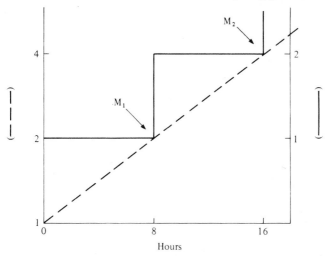

Fig. 37. A model of mitotic cycling in *Physarum* (relaxation cycle). A cytoplasmic factor (dashed line) accumulates continuously until a critical value, relative to some nuclear parameter, is reached (solid line) which initiates mitoses (M_1, M_2): hour glass model. Alternatively, a cytoplasmic factor accumulates continuously until a given number of nuclear receptor sites (solid line), postulated to double immediately after nuclear division (arrows), become saturated: nuclear-site titration model.

maier, 1978). It is actually a rephrasing of the classical nucleo-cytoplasmic ratio concept put forward at the turn of this century, with some ramifications to explain various perturbation experiments. It assumes that some state (or factor(s)) slowly accumulates to a critical value which is quickly adjusted at mitosis to a low basic value, and then the sequence starts all over again. An oscillator is postulated that builds up 'tension' which is quickly 'relaxed'; hence the model is also called an 'extensive relaxation–oscillator model'.

In its current state an 'activator', which must be a labile molecule (see above), accumulates continuously in the cytoplasm, moves to the nuclei and titrates 'nuclear sites'. The activator could be something like the elusive division protein postulated by Zeuthen (Zeuthen & Williams, 1969). When all sites at or in the nucleus are covered, mitosis is triggered and the 'relaxation' quickly comes about, as the nuclear sites double immediately after mitosis and then have to be titrated again and again (Fig. 37).

This is a logical story, but looking over the facts it reminds one of a situation known as 'counting angels', which assumes that there are angels in the first place. There is no evidence for an 'activator'; there may well be 100 different factors specifically involved in the mitotic-apparatus cycle. Moreover, there is no trace of nuclear sites that double after mitosis. Here, one might speculate that – if the master initiation of S phase turns out to exist – all origins of DNA replicons double directly at telophase. This would indeed provide a good and testable hypothesis for assessing DNA replication initiation and why it starts only once per cycle. However, it would not explain nuclear division.

In other words, the model is based on two assumptions. It is useful in describing the cell cycle, but it really remains a black box which may turn out to be neither black nor square. As long as we know that the weakest of low-frequency electromagnetic waves influence the mitotic clock of *Physarum*, we had better be prepared for completely unexpected control mechanisms.

This brings up another model which is basically different. It requires good and, for a trained mathematician, simple equations and no experimental data whatsoever. As far as the hour glass model is concerned, it is required that mitosis be part of the mitotic clock. Figuratively speaking, at mitosis the sand has run out, and the glass has to be turned over to reset the clock. However, the so-called 'limit-cycle model' assumes a different situation. The cell contains an oscillator, a cycle of chemical reactions, that runs independent of mitosis and continuously. Hence, mitosis happens to be triggered if a certain mixture of chemicals has been reached. In its simplest form, two substances, call them inhibitor and activator, interact with each other by auto-catalysis and cross-catalysis in some non-linear fashion, with the result that a continuous oscillation in the actual concentration of the two substances takes place. In one graphical representation this looks like a cycle (Fig. 38*a*) and its limits are set by the kinetics of the chemical reactions. Such oscillations have been observed in some inorganic solvents and in glycolysis (at least, *in vitro*; Hess & Boiteux, 1971). However, this model is more powerful; it explains almost everything else in biological development, from evolution, through the mitotic cycle and embryonic patterns to morphogenesis. In addition, the model cannot be tested, as it does not predict experiments. However, it can rule out certain other models. To fit the limit cycle to mitosis of *Physarum*, one has to set a certain value (say 2*x*/1*y*, i.e. twice as much activator as inhibitor) and expect that mitosis is triggered. This is shown simplified in Fig. 38*b* and has been elaborated in much more detail by Kauffman & Wille (1975). Of course, although they may exist, there is no evidence for the activator or the inhibitor. Similarly with those two parameters, set against distance and space instead of time, one can explain the segmentation of a worm or the whole body plan of an insect. One expectation for *Physarum*'s mitotic cycle is implicit in the graph: if the activator does not reach the required level, mitosis must be skipped altogether and occur a full cycle later. The many fusion experiments with growing *Physarum* plasmodia have never shown that. For a detailed discussion of *Physarum* and the limit cycle the knowledgeable discussion by Tyson (Tyson & Sachsenmaier, 1978) is recommended.

However, a starved plasmodium extends its cycle, it seems, in big jumps, to 24 and 38 h as shown painstakingly by the Guttes (Guttes *et al.*, 1969).

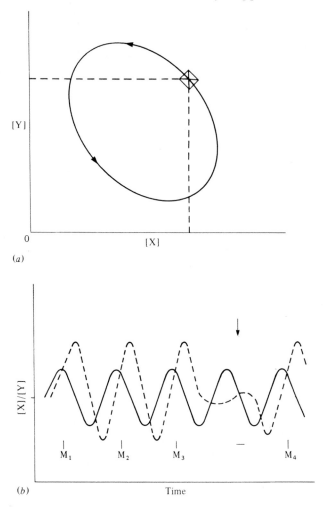

(a)

(b)   Time

Fig. 38. Another model of endogeneous cycling in *Physarum* (limit cycle). (*a*) The concentrations of two substances, X and Y, change with time, and mitosis (×) is triggered when a certain ratio is reached. (*b*) Proposed consequence of a pertur-bation (arrow) preventing the critical ratio of X : Y.

The limit cycle mathematics could be fitted to that observation, but only direct analysis has given us a hint of how *Physarum* does it. It eliminates a lot of nuclei, maybe to keep pace with the slow, or actually the negative, growth during starvation; thus the nuclear/cytoplasm relationship is maintained.

Another comment may be allowed concerning the now already classic 'cell cycle concept', originally proposed by Howard & Pelc (1953). In the fifties this led to a very valuable concept, as it marked a successful cross-fertilization between cytologists and biochemists. The cell cycle arose by superimposing two discontinuous events in the life of a cell: nuclear division and S phase. The G_1 and G_2 phases have been formal definitions of cell-cycle segments, meaning temporal 'gaps' before and after S phase. These gaps have been filled with various and many data (Mitchison, 1971; Prescott, 1976), which up till now have not shed any light on the cell-cycle concept. There are numerous cases in which one or the other gap is missing; in protozoa or in embryos, for example. Clearly, if there are specific events required to prepare DNA for replication, they may occur after mitosis, thus defining a G_1 phase, or before mitosis as part of G_2 phase. Such events might just take a short time in a rich environment, or a long time in a poor one, which would define a G_1 phase in the latter case. What we need to understand first is the events as such, not so much when they occur and how long they last. It may well be that all cases in which mitosis and DNA replication are uncoupled in a differentiated metazoon are proof that this particular cell is no longer in balanced growth but has embarked on a differentiation pathway. In such cell a gap phase might have become established as a loss of a growth function or a gain of a differentiated state by some sort of heritable alteration, perhaps even a somatic mutation in its widest sense, to enact the phenomenon of 'determination'. This is an extreme view, just to show that new questions can be asked if we give up the classical concept. The new question is: why is there a gap phase when many proliferating protozoans and uncommitted cells can do without it?

The mobile plasmodium

In addition to autonomous mitotic divisions, there is another endogeneous oscillation.

The shuttle: protoplasmic streaming

The protoplasmic streaming in *Physarum* is really remarkable, not only is it a fascinating phenomenon and a 'must' for any student of cell biology, it is also amenable to convenient studies. Many experiments have been done by a large number of researchers, but communication among the various groups interested in the cell cycle has been poor (for reviews see Berridge & Rapp, 1979; Wohlfarth-Bottermann, 1979; Nachmias, 1979*a,b*; Hatano *et al.*, 1979; Pepe *et al.*, 1979; Kessler *et al.*, 1980).

As the name implies, the protoplasm gets moved in one direction for a short while, stops, then turns back for a while, and so on. The period is quite stable and varies only between 1.5 and 3 min; the speed of flow is really enormous for any cytoplasm, up to 1 mm/s. Not the whole plasmodium shuttles, as there are two distinct compartments: the endoplasm which shuttles and the ectoplasm which is more solid and makes up the veins in which the endoplasm flows.

The analytical work has been done either on small intact plasmodia or on excised plasmodial strands. A classical experiment was done by Kamiya (1959). He put a plasmodium in his 'double chamber' (Fig. 39*a*) and applied just enough pressure to one half of the chamber, to block the streaming of the endoplasm into that half of the mould. Thereby he could measure and define the 'motive force'. The outcome of this elegant experiment was surprising. Even if the flow was stopped, the motive force varied, thus unveiling a contraction rhythm independent of the flowing endoplasm.

More direct evidence for the contraction of the ectoplasm-tube as the cause of the streaming comes from a very simple experiment. If the endoplasm is blown out of the tube, and the tube of ectoplasm is attached to a tension meter, the contractile rhythms persist for quite some time (Fig. 39*b*). Test solutions can be substituted for the endoplasm and one learns that energy in the form of ATP (optimum conc: 0.2 mM) and calcium (at least 10^{-7} M) are required for this *Physarum* 'muscle' to work.

With this knowledge we can forget for a while whether the endoplasm gets pulled or pushed, as its involvement is passive anyway. We can learn more about the contractile apparatus if we compare the endoplasm and ectoplasm in structure and chemistry. Observations on a living plasmodium have demonstrated that the outer plasmalemma is very mobile, and that about 80% of it is involuted, forming a labyrinth of invaginations from which vesicles are pinched off on the inside, and other vesicles are fused into it as they get shoved (like the slime) out of the plasmodium. There are contractile vacuoles and other kinds of vacuoles, too.

The depth of the invaginations defines the thickness of the tube or ring of ectoplasm (Fig. 40*a*). It contains bundles of microfibrils which are attached to the invaginations and arranged in orderly fashion. There is also a felt-like network of thin filaments covering the inside of the plasmalemma, the cortex. No filaments are in the normal endoplasm, the site of most nuclei, however there are some nuclei that have become embedded in the ectoplasm gel.

The bundles of fibres (microfibrils or microfilaments) are present *in vivo*, as they show up in the polarising microscope, but they change in phase with the contraction of the tube. The contraction shortens the tube and makes it thinner (Fig. 40*b*). Fixation at distinct intervals during contraction and relaxation shows that the bundles appear at contraction, and most of them disappear again during relaxation, providing an impressive and fast intracellular case of morphogenesis. The bundles are arranged longitudinally and are fixed at the plasma invaginations; they also occur as a ring or spiral around the tube. It is tempting to assume that these bundles are the contractile

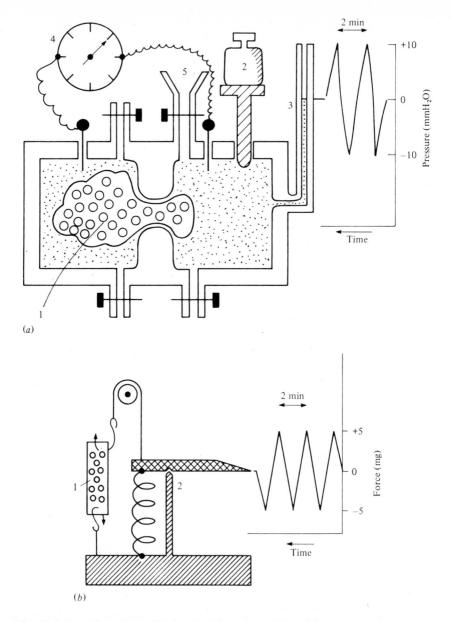

(a)

(b)

Fig. 39. Measuring plasmodial forces. (*a*) A plasmodium (1) is prevented, by pressure exerted on one chamber (2), from migrating into that chamber. It causes the fluid in the attached manometer to oscillate regularly by ±10 mm H₂O (3). The pressure needed to hold the plasmodium back defines the motive force. Change in potential in the two chambers can be monitored (4), and effect on motive force can be analysed by changing the environment in one chamber (5). (*b*) A plasmodial strand (1) is hooked to a tension meter (2), and the tension force (approx. ±5 mg) can be analysed.

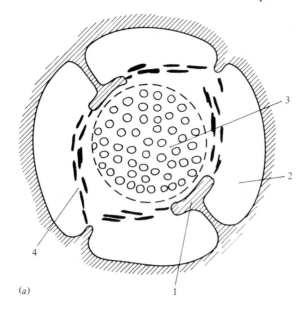

(a)

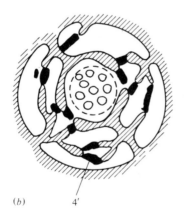

(b)   4'

Fig. 40. A schematic view of the contraction of a plasmodial strand: (a) cross section relaxed; (b) cross section contracted. Plasmalemma invaginations (in conjunction with the cortex) (1) delineate ectoplasm (2) from endoplasm (3). Microfilament bundles, attached at the involuted plasmalemma (4), shorten (4') making the strand thinner and displacing the ectoplasm.

apparatus which appears periodically as required to contract the ectoplasmic tube; it also looks like a smooth muscle while it contracts. Therefore, the microfibril bundles and their dissociated components may be considered the motile system of an unspecialised cytoplasm of a very old creature.

Actomyosin has been demonstrated in the plasmodium (after glycerination) and in plasmodium extracts. The microfilaments have been identified *in situ*, as they bind heavy meromyosin obtained from a rabbit. After this treatment they appear to be decorated by arrow-heads; the accepted interpretation of this pattern is that the microfilaments are oriented both ways. Actin has been identified: it is a molecule of 44 000 daltons, is very abundant in the cytoplasm, and makes up 5–20% of total protein. Its amino-acid sequence is known and is only 4% different from that of the mammalian cytoplasm (Vandekerckhove & Weber, 1978), and even more similar to *Dictyostelium* actins. The composition of the respective genes may hold far-reaching information on the evolution of a major structural gene in eukaryotes. Actin can assume either monomeric form (G-actin) or exist as a filament (F-actin). This can be demonstrated *in vitro* and 5′-AMP prevents F-actin formation. There is enough actin present in the endoplasm to form filaments by polymerisation, but this does not occur naturally. However, if a plasmodial strand is stretched in length, actin filaments appear inside the endoplasm, as the tension force increases. After the ensuing contraction of the whole strand the fibrils disappear again. This definitely shows, that the monomers of actin are present in the endoplasm.

In another approach, a vein is punctured and a drop of endoplasm exudes from the wound. Vesicles flow to the border between the endoplasm and the surrounding liquid; they fuse quickly (within seconds) forming a membrane sheet around the drop. Then the drop differentiates into ectoplasm and endoplasm over a period of about ten minutes, and as soon as microfibrils appear, the whole drop gets pushed back into the mainstream of the endoplasm. Some people have wondered whether or not these swift ultrastructural changes can be classified as 'differentiation'.

Another protein, actinin, with the same molecular mass as actin has been purified; it prevents polymerisation of G-actin to F-actin. It is very abundant (5% of total protein), and it may be involved in regulating the G-actin-state inside the vein, but also in dismantling F-actin in the filament bundles after each contraction of the tube. At least some of the F-actin must be degraded during relaxation, as the drug phalloidin, which freezes the F-actin state, blocks cytoplasmic streaming after injection and the microfibril bundles get bigger.

No doubt, the 5–7 nm actin filaments are essential in contraction. How about the other essential partner for contraction, the myosin? Myosin has been identified as a heavy molecule of 5×10^6 daltons and two light chains. Its content is much less (0.8% of total protein) than in striated muscle cells, but similar to the concentration in the cytoplasm of other cells. Its proportion relative to actin in an actomyosin preparation is quite low (1 : 200). Its state of polymerisation into thick fibres is not clear. After glycerination 10–12 nm thick fibres of 0.5 μm length have been observed, and magnesium or calcium ions facilitate aggregation into large fibres with heads and tails *in vitro*.

Although not completely proven, the actomyosin of *Physarum* may look like this: bundles of F-actin and oligomeres of myosin, both in antipolar orientation and fixed at the plasma membrane, produce tension by a sliding mechanism which would bring about the contraction of the vein; this structure closely resembles a smooth muscle. While this is how it looks, how does it work? Actomyosin in solution, or the whole plasmodium (after glycerination) contracts upon the addition of ATP. Obviously, a Mg-dependent ATPase has become activated. *In vivo*, the contraction requires calcium, and an oscillation in the free Ca^{2+} level could explain rhythmic contractions. Before we ask whether this important ion oscillates, we can ask how it influences contraction. It seems to act in two ways. Calcium has a high affinity for one of the light chains of myosin, and it may effect the rate of phosphorylation of that protein. The latter process is known from smooth-muscle contraction and involves a regulatory protein, which is present in all cells, called calmodulin. This ubiquitous calcium receptor (Cheung, 1980), a protein that has been identified in *Physarum* (Kuznicki & Drabikowski, 1979), probably activates the kinase which phosphorylates the light myosin chain (LC II), which in turn activates the actomyosin ATPase which finally causes the contraction.

An increase in free Ca^{2+} just at contraction has been concluded from oscillating light-emission from injected equorin (a probe that emits light when enough free Ca^{2+} is present), and we have to ask where the calcium goes when the luminescent light goes out. First, one could suspect that the changes in light intensity may just reflect variable thickness of the plasmodium, as it contracts, and the conclusions from these studies might arise from an artifact. However, histochemical studies have also located free Ca^{2+} in the cytoplasm during contraction in the cytoplasm.

At any rate, there are efficient ways to get rid of calcium from the cytoplasm by three routes. There must be, since outside concentrations of calcium are 1000-fold higher than those required inside for contraction. One mechanism is to pump calcium across the cell membrane; another is to pump it into vesicles that lie close to the site of action. (In striated muscle, the sarcoplasmic reticulum has an extra pump, a Ca,Mg-ATPase). In *Physarum*, one class of vacuoles has been identified which accumulate calcium *in vitro*, and a Mg-ATPase of about 100 000 daltons has been demonstrated in the vesicle membrane. This enzyme cross-reacts with an antibody to the respective enzyme obtained from the sarcoplasmic reticulum of the rabbit – it seems to be another conserved protein. The third calcium store in *Physarum* is the mitochondrion. Accumulation of calcium occurs by an elaborate process that seems to be too slow to operate effectively during the contraction oscillations.

The next question is, where does the calcium come from once every two minutes or so? That, actually, is the crucial question, as it may lead to the identification of an 'oscillator'.

When the plasmodium is kept free of external calcium (below 10^{-8} M), the contraction rhythm is not disturbed for about one hour. Therefore, the uptake of calcium does not seem to be crucial. Other ion fluxes through the outer membrane, indirectly affecting the interior calcium level, do not seem to be involved either. This can be assumed from two observations. Specific inhibitors of the proton pump and K,Na-ATPase in the outer membrane (ouabain and valinomycin) do not influence the contraction rhythm. On the other hand, a general metabolic oscillator (such as glycolysis) does not seem to be involved either, as anaerobiosis has no effect, nor does a blockade of glycolysis by iodoacetate disturb the contraction rhythm. Nor is phospho-fructokinase regulated by an allosteric mechanism. This is similar to the situation in *Dictyostelium*, but unlike the yeast system (G. Wegener & H. W. Sauer, unpublished). On the other hand, slight fluctuations in the membrane potential, indicative of ion fluxes, go on even if the protoplasm does not stream. Therefore, it seems that the release and uptake of calcium is an endogeneous process, like in striated muscle. There it is controlled by membrane depolarisation and the action of nerve cells. How might it be controlled in *Physarum*?

At this point in the discussion it is time for cyclic AMP (cAMP) to step in and offer a solution to our problem in the form of yet another model (Rapp & Berridge, 1977). A high calcium level means more calcium–calmodulin complex. This receptor molecule activates not only actomyosin-ATPase but adenylate cyclase also. Therefore, the cAMP level goes up, which in turn activates a kinase that phosphorylates the calcium transport ATPase in the calcium vesicle membrane. Consequently, the concentration of free Ca^{2+} is lowered. That means less calmodulin is bound, and the cAMP level goes down as the pump slows down. This in turn causes the calcium level in the cytoplasm to rise again, and so on. No conclusive evidence to support this model is available for *Physarum*, and some of the data are conflicting (Achenbach & Wohlfarth-Bottermann, 1980).

We are left with a model, but we return now to the intact plasmodium, where another very fascinating observation has been made: the whole plasmodium of over 5 cm in diameter contracts in unison. The shuttle streaming is not a peristaltic wave but a mono-rhythmic phenomenon. This striking observation comes from a very ingenious and simple application of film analysis to a whole plasmodium (Grębecki & Cieślawska, 1978). When one looks at the film, the shuttle streaming and the contractions of the veins do not seem to be well coordinated at all. However, if the film is projected again and again, and each time a different spot along a vein is analysed, as it gets dark and light in phase with the constriction rhythm (which is documented by exposing a film strip that is moved behind the projector, slowly), overall mono-rhythmic contraction has clearly been revealed. This raises the question, how does this gigantic 'cell' coordinate such an accurate rhythm? Does it have some sort of nervous system in the ectoplasm, and how does it work? We can assume from the experiments with the poisoned proton pump, that the outer membrane is not involved.

If one strand in a network is gently touched with a blunt needle, that strand stops shuttling for a while, but only at that local area. On the other hand, if such a strand is stretched out considerably, it stops shuttling. When the contractions start again, it can be seen that local areas over the whole length of the strand oscillate independently of one another, but they become resynchronised within about half an hour. In still other experiments it can be shown that the streaming endoplasm is required to maintain or achieve synchrony of the contraction rhythm. This has been shown in the intact plasmodium and in a fusion experiment, the former being another application of the double-chamber method of Kamiya. When the pressure has been applied to prevent the endoplasm streaming from one half to the other half of the plasmodium, then both halves oscillate in a different contraction rhythm. If the pressure is released and streaming is permitted again, it takes only 5 min for both halves to oscillate in synchrony again. In the fusion experiment, one strand showing synchrony in contractions begins independent oscillations in two halves after it is cut in the middle. If those two pieces are fused by placing another piece of strand in between them, the whole complex oscillates in synchrony again. One possible explanation for these results is that the many endogeneous local oscillators all become synchronised by the stretch that the ectoplasmic tube experiences while the endoplasm is pushed through the tube. Such a model of 'stretch-entrainment' has been proposed for some smooth muscles. If the stress or stretch is relieved by applying a drug that interferes with the cortex lining of the tube (like cytochalasin), the strand gets ruptured, and endoplasm flows out. As expected, the oscillator becomes interrupted locally until the lesion has healed and the tube has become sealed off again.

This observation finally brings us to another biological implication of the rhythmic contractions, aside from distributing protoplasmic contents more or less evenly. Shuttle streaming brings about morphogenesis of the plasmodium and is the motor, although not the pilot, for locomotion of the whole mould. The main point here is that at each contraction cycle, a conversion of endoplasm to ectoplasm occurs wherever space is available. We can understand how a macroplasmodium remains stationary and grows as a completely round disc. As the whole structure compresses, the endoplasm gets pushed and squeezed out into the periphery, and as the contraction of the ectoplasm is relaxed, the endoplasm gets sucked back. At the periphery the plasmalemma gets extended, i.e. stretched out, which causes this area to contract and aid in pushing back the endoplasm. It has been estimated that about 30% of the endoplasm gets converted into ectoplasm each cycle. This describes correctly that the growth of the plasmodium occurs only at the edges; however, it does not explain it. It is also conceivable that the outer membrane is more pliable, because it is 'smelling' something beyond its reach. Therefore, it could still be 'pushed' or 'pulled' after all. As long as the capacity to synthesise cytoplasm allows, a round expanding disc results. When more ectoplasm gets made at the front than can be provided from inside (by de-novo synthesis of cytoplasm), the disc breaks up in the centre, and finally the whole creature moves away (Fig. 2b).

The occurrence of a quite regular reticulated net from a flat homogeneous disc is a fine example of pattern of formation, something reminiscent of the elaborate veins in plant leaves; yet this has not been studied any further in *Physarum*.

Before we turn to migration of the plasmodium, let us note three facts that ought to be considered for the other endogeneous rhythmic and synchronous event: nuclear division. Although the plasmodium as an entity is completely in balance, it is not at all homogeneous: there is a gradient in age from the old centre to the young periphery. The nuclei are tossed around the whole endoplasm but they consist of a mixture of old and young nuclei and those embedded in the thick and more stationary ectoplasm. In addition, the expansion of the growing sheet shows an abrupt stop at mitosis, and then it goes on growing after a few minutes (Miller & Anderson, 1966). It seems conceivable that the halt in expansion of the plasmodium, as well as the stopping (or slowing) of the shuttle at mitosis (which correlates with the highest pH value), is the consequence of a general stiffening of the whole protoplasm, maybe a mechanism for protecting the nuclei from colliding and giving them something to hold on to during their division process, after all. As long as the natural mitotic synchrony in a big plasmodium can 'only' be accurately determined within a period of ten minutes, and in view of the fact that the contraction oscillations and their control mechanisms occur within minutes or possibly seconds, it will be very interesting to find out, if the nuclei in the ectoplasm divide a little earlier than those in the endoplasm, and whether the halt in growth correlates tightly with nuclear division. If it does, does it stop before (because of) or after (as a consequence of) mitosis?

Mould on the move

So far we have considered the structure and function of the motile apparatus mainly as a convenient 'model system' of intracellular communication.

It is a contraction oscillator, some sort of smooth muscle, and an extreme case of rapid protoplasmic streaming, as well as a case of fast intracellular morphogenesis.

Model studies have used a strand cut out of a large, vigorous plasmodium, or microplasmodia submersed in liquid culture medium, or the macroplasmodium-disc which arises by simultaneous fusion of microplasmodia. Certainly, none of these conditions is very natural to *Physarum*. Many of the pertinent observations on migration have been done by Lewis (1942) who described how a small plasmodium moved over the substrate (despite the rapid shuttle streaming of 1 mm/s) just like a true amoeba, or myxamoeba, and at about the same speed of 5 mm/h. As it does so, it takes on a definite form, and it looks like a fan (Fig. 41a). Taxonomists refer to this form as a phaneroplasmodium, typical for Physarales among the acellular slime moulds (Alexopoulos, 1960). The handle of the fan is the tail, and at the front is a thin sheet. A tube of ectoplasm surrounds the tail and the body, and forms the frontal sheet, leaving a vein of endoplasm inside. The tube is tapered at the tail end and branches out at the front. The creature creeps along the surface over a slime cushion which it exudes as it moves, leaving a trail behind it.

What happens when the whole structure contracts radially and longitudinally? It gets thinner and shorter (Fig. 41b–d). As the layer of ectoplasm is thicker in the tail end, the shortening is more effective there. As this happens, the endoplasm gushes into the head, since the tail is rigid and stuck in the slime, and the plasmalemma invaginations are pushed out, thus producing new pseudopodia. But that is only part of the story. At the rear of the plasmodium, a conversion of ectoplasm into endoplasm takes place which recycles some of the motile apparatus in its disassembled pieces of G-actin and myosin monomeres which end up in the head, first as a net increase in endoplasm, later as part of the new ectoplasm sheet. The head is the youngest part of the plasmodium, and as long as overall synthesis exceeds degradation, the whole fan gets bigger as it migrates.

Now let us consider the relaxation phase of the contraction cycle. The plasmodium gets thicker, and it expands in length. At the rear end it cannot revert to its former length because some of the tail has been dismantled. The endoplasm moves back: it gets sucked back from the front at the same time as the pressure that has been caused by the contraction of the posterior is released. Added pressure is applied by those actomyosin fibres that now appear in the front sheet, probably due to stretch-entrainment. However, some material stays at the front as newly-created ectoplasm. As contraction and ectoplasm–endoplasm conversion are not synchronous, the movement is a continuous process that is further smoothed out by the slime secretion.

Now let us consider a very well-nourished plasmodium: much more endoplasm will get into the front as more is synthesised. The head becomes

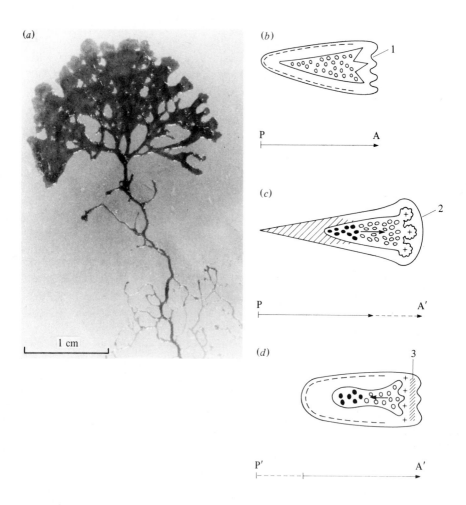

Fig. 41. (*a*) Phaneroplasmodium. (*b–d*) A plasmodium taking a step. (*b*) The polarised plasmodium has an involuted plasma membrane with a less prominent cortex (broken line) at the front (A), (1), and contains endoplasm (open circles). (*c*) Microfilament bundles appear and contract in the posterior half (P) (parallel lines), where some ectoplasm is converted into endoplasm (closed circles) and pushed forwards (arrow), thus adding new ectoplasm (crosses) at the front, and straightening out the plasmalemma membrane (2). (*d*) The microfilaments disappear at the posterior end, and some new microfilaments now appear at the frontal lobe (3), as the remaining endoplasm is pushed back towards the rigid tail (arrow).

bigger and bigger, and the tail gets absorbed. Consequently, the plasmodium becomes a disc, consisting only of the head part. The bipolarity has been lost, and further growth of this structure occurs centrifugally. The plasmodium stops migrating altogether. If such conditions are met in the laboratory in a liquid medium, the plasmodium becomes suspended and is no longer attached to a solid substratum. Then it grows as a sphere, like microplasmodia in a shake flask. The faster the flasks are shaken, the smaller the spheres become, as they break apart when they grow too big. They are also not allowed to fuse, maybe because they are continually agitated, and they do not have a tendency to fuse anyway if they are well nourished. (This is why microplasmodia are starved for a while if a sizeable macroplasmodium is required in the laboratory.)

What happens if a moving phaneroplasmodium uses up more material in the endoplasm–ectoplasm transition process at the front than it produces? It gets smaller, but it also migrates faster (3–5 cm/h). The shuttle of the endoplasm becomes more vigorous, although the frequency of the oscillator is not altered. As streaming of the endoplasm is passive, the contraction force of the ectoplasm tube must have increased. Indeed, the amplitude of the contraction becomes greater and, in addition, the actin concentration reaches up to 20% of the total protein. This alone, however, cannot account for the more intensive migration. If the degree of dismantling of the tail remains constant in a starving plasmodium, it will not move faster than before. But, if the plasmalemma at the front gets softer, even without an increase in motive force, migration must speed up due to the longer strides the plasmodium will now take. As a speculation we can propose that a rapid utilisation of the old ectoplasm at the tail, together with a change in anterior plasmalemma rigidity enables a plasmodium to move faster.

There are two ways in which a change in behaviour in any organism can be brought about: from forces within the organism, or from forces acting on the organism from the outside. We know very well that the mitotic and the contraction oscillations are endogeneous processes, and the vigorous migration of a hungry plasmodium may indicate that it is uneasy or generally in a poor condition. Under such circumstances it may get ready to do something quite different. Developmental biologists have used the term 'competence' and the behaviourists 'appetence' for this. The phenomenon is widespread and very important for any living being, yet it is not understood.

Therefore, if we deliberately disturb the plasmodium from the outside, we may learn something about what is normally going on inside it. So, how can we force the plasmodium to go somewhere? As we have seen already, a fat sedentary individual starts moving only if it gets short of food. For this general question we have provided here a general answer. In the following chapter we shall try to obtain a more specific answer.

But we might add that the current views of amoeboid movement as reviewed by Taylor & Condeelis (1979), and the classical observations on *Physarum* plasmodia by Lewis (1942), plus the recent work on migration in *Physarum*, make up a general and unified picture which favours some sort of 'pushing'; just a little extra endoplasm goes into the pseudopodia than comes out again. This is quite different from the concept of mammalian cell movement, in which the highly structured whole cell, with the microtubules (MT) and the intermediate-size filaments (IF) in addition to actomyosin, seems to pull itself forward after the filopodia of the frontal lamella stretch out, search for an attachment site, and there attach firmly for a while, until the next movement. However, as we shall now see, the front end of a *Physarum* plasmodium is involved in directive migration, too.

One final comment on the shuttle: in every case we have considered a shuttling plasmodium. An eminent question for the developmental biologist has not been addressed: why does the microscopic amoeba move without shuttling, and how does the shuttle develop in the newly born plasmodium?

The sensitive mould: chemotaxis

Let us begin by discussing an unphysiological experiment: the application of an electric field across a plasmodium. The plasmodium will start moving to one pole: the cathode. However, we do not know whether it moves *towards* the cathode or *away* from the anode. This behaviour is not some sort of elec-trophoresis, as might be suspected from the overall negative charge of the plasmodium surface (the resting membrane potential in *Physarum* has been measured as about -25 mV), but it is an active migration during which pot-assium ions are extruded from the pseudopodia at the moving front. This loss may be balanced by an influx of calcium ions. While this is not clear, we can certainly conclude that the membrane is involved in this directional flow (and we should remember that this seemed not to be required for the endogeneous contraction oscillations).

Local changes in ion fluxes have likewise been observed in protozoa, and have been shown to inform an egg of the famous brown alga *Fucus* where to make its root, in other words, how to achieve 'polarisation' (Jaffe, Robinson & Nuccitelli, 1975).

Under normal conditions, directed cell movements play a pivotal role in the morphogenesis of animal embryos in two ways: the cells respond to stimuli from other cells, or to extracellular signals. In the adult body, white blood cells circulate and watch out for foreign material which they recognise as 'foreign' and devour. Therefore, signalling and recognition, as well as cellular migration, are basic requirements for coordination in protozoa, and in metazoan organisation and development (Monroy & Moscona, 1980).

A very elegant example of cell communication is the 'social' amoeba of

Dictyostelium (Gerisch, Malchow & Hess, 1974). When the amoebae are hungry (or have become 'competent') they are very sensitive to an external signal, an acrasin, which in *D. discordeum* is cyclic AMP (cAMP). This is secreted rhythmically from one very hungry amoeba: the leader. These pulses are the signal which causes the surrounding amoebae to react in two ways. Firstly, they throw out pseudopodia towards the leader and thus approach it gradually. Secondly, they produce a lot of cAMP themselves, and while they contract they spread it around, thus amplifying the signal radially to a community of about 10 000 amoebae. All of them migrate in rhythmical procession towards the centre and form a pseudo-multicellular organisation, the slug. Much more detailed knowledge of the developmental sequence has been accumulated using sophisticated biophysical, genetic, and immunological methods. Thus, a brief pulse (15 s) of cGMP in response to the signal-reception and the appearance of a polar positioning of specific surface antigen (specific for the aggregation stage) have been detected. Here suffice it to say that rhythmical contractions (with a period of about 0.5 min) are also involved in morphogenesis, just as in the formation and transformation of the phaneroplasmodium of *Physarum*.

Also we have learned to be cautious in drawing comparisons with other organisms. A very close relative to *Dictyostelium* which looks just like it, *Polysphondylium*, is completely insensitive to the ubiquitous cAMP and signals with the help of a small polypeptide; also folic acid can substitute for cAMP in *Dictyostelium*.

However, in view of the social-life period in *Dictyostelium*, we should ask whether chemotaxis plays a role in communicating among different plasmodia of *Physarum*. An elegant experiment has been done in China (Tso & Wong, 1978). Two plasmodia were put on a porous membrane, one on either side. The membrane was suspended in a moist chamber and the migration pattern of the two plasmodia was investigated. The result was that they took no notice of each other. Therefore, if plasmodia fuse when they come together, they seem to do so unintentionally. This does not mean that they are insensitive, since they can sense light, a number of chemical substances, as well as moisture and temperature, and can respond by either moving away or by being attracted. If chemicals are the signal, they move chemotactically by definition.

Chemotaxis in *Physarum* can be very conveniently studied at the behavioural level (Carlile, 1970; Knowles & Carlile, 1978). If a hungry plasmodium is placed between two strips of non-nutrient agar (Fig. 42), it will migrate either to the right side or to the left. The probability for random migration going to the right is 0.5. To test whether a substance can be recognised, it is included in one agar strip, from which it diffuses towards the plasmodia to be tested. If they like it well, they will go to the right (probability 1), if they dislike it, they will go to the left (probability 0). With this assay it

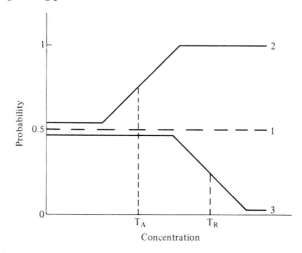

Fig. 42. Approaches to analyse chemotaxis in *Physarum*. A hungry plasmodium, placed between two pieces of agar, will move either to one piece or the other. If both pieces are identical, the plasmodium will move to one or the other side with equal probability ($P = 0.5$) (curve 1). For an attractant on one side, probability to go there increases (curve 2); for a repellent, probability decreases (curve 3) – thus thresholds (T_A or T_R) are defined.

has been found that certain carbohydrates that the plasmodium can metabolise are attractants, some are not. Glucose attracts above 0.1 mM concentration; deoxyglucose is also attractive, although it is not utilised. Sucrose and fructose are repellents. This assay has been applied to many substances such as amino acids, nucleotides and various salts (Chet, Naveh & Henis, 1977; Kincaid & Mansour, 1978). For the plasmodium it is now clear that a good correlation between chemotaxis towards organic molecules and their nutritional value does not exist. However, an easy bio-assay to detect threshold levels for attractants and repellents within a few hours is available.

For an understanding of directional movement, we need to know how the respective signals are sensed, then transduced, and finally, how they affect motility. The classical double-chamber experiments have been ingeniously modified to measure the effect of chemicals on the motive force (Fig. 43), and to find out whether it gets stronger or weaker (Ueda & Kobatake, 1978). Remember, motive force was the pressure put on one chamber to suppress endoplasmic flow (p. 104). Many test substances in the other chamber influence the flow pressure, and a general rule has been recognised: attractants increase the force, repellents decrease it (the differences are equal in both cases, either $+10$ mmH$_2$O or -10 mmH$_2$O in the manometer). The thresholds are the same as those determined in the behavioural assay, but

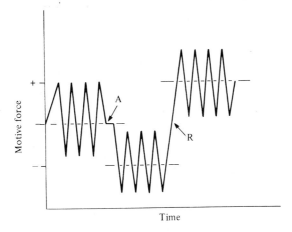

Fig. 43. Effect of an attractant (A) or repellent (R) on the motive force, as determined by the double chamber method.

there is no correlation between the degree of change in motive force and the concentration of the substance above the threshold. It is an all-or-nothing effect. The motive force can be modulated by the frequency of the contractions and by the amount of ATP available. No good correlation with the periods of oscillation has been detected, although some contractants speed up the contraction cycle (which cannot become shorter than 1.5 min in liquid or 1 min in the air). No effect on the endogenous ATP level has been noted with the attractants, whereas some repellents increase the endogenous ATP level by as much as a factor of two.

A more direct assay of the force production by the contraction oscillator is the change in tension of an isolated strand suspended in a liquid containing the test substance, measuring the effect with a tension meter. Again a general rule emerges; attractants release the tension, repellents increase it. Moreover, these changes do correlate with the concentration of the chemotactically active substance.

None of these measurements tells us how the signal is sensed but, once again, the double chamber has given us a clue (Ueda & Kobatake, 1978). A membrane potential difference can be detected just by connecting the two chambers, if a chemotactically active substance such as glucose is added to one chamber. The results of such measurements, which have been substantiated in more tedious analyses of the transmembrane potential (inserting one electrode into the endoplasm), indicate a partial depolarisation from -50 mV to about -45 mV (Hato, Ueda, Kurihara & Kobatake, 1975). This observation can be made for any chemical that has elicited a response in the behavioural test, provided it is above the threshold value, but irrespective of whether it is an attractant or a repellent.

While these observations could also be explained by a change in membrane permeability, it has been clearly demonstrated that the outer membrane becomes altered on chemoreception. This was done by watching small plasmodia migrate electrophoretically in an electric field and measuring the effect of various chemicals on the speed. From this, the zeta potential, or the surface charge, was estimated as −50 mV and again a monotonous change of 5 mV (to about −45 mV). This might indicate that about 10% of the negative groups at the outer surface of the plasmalemma get buried in the membrane upon chemoreception. Therefore, a conformational change of the plasmalemma could be involved in chemoreception. This has also been shown with a model compound called anilino-naphthaline-sulphonate (ANS). The magnesium and sodium salts of this material fluoresce when they are incorporated into the membrane. The wavelength of fluorescence shifts from about 500 to 450 nm as the membrane becomes more hydrophobic. Relevant tests were done with a white mutant of *Physarum* (Ueda & Kobatake, 1979). The spectral shift occurred only if the ANS was added to the plasmodium above the threshold. At that moment the membrane also became more fluid, as concluded from fluorescence-polarisation with incorporated ANS. Consequently, the membrane gets stiffer and more hydrophobic when sensitised. Yet another line of supportive evidence can be cited: with a series of alcohols a good correlation of threshold and chain length, or sensitivity with hydrophobicity, has been determined (Ueda & Kobatake, 1977).

These and other biophysical studies have led to the conclusion that the primary site of reception is just above the membrane surface and involves a disturbance of the surrounding water molecules, i.e. the 'water structure'. Glucose, an attractant, seems to increase the rigidity of the water cover (Terayama, Ueda, Kurihara & Kobatake, 1977).

In addition to all the sophisticated and complex measurements, a very conclusive role for localised changes at the cell surface comes from a simple determination of the adhesive force at the moving front of a plasmodium. This is measured as the force that is required to pull off a plasmodium that has been put on a hydrophobic surface, such as a simple plastic Petri dish. It has been shown that attractants increase and repellents decrease the adhesive force.

While these results strengthen the view of the importance of the outer surface of the plasmodium, the inner surface seems to be equally involved in directed migration. The drug cytochalasin A causes 'blebbing' of the surface, into structures which resemble pseudopodia. This happens particularly at the moving front. The drug is assumed to interfere with the felt-like network making up the cortex layer underneath the plasmalemma. As a last argument, it has been shown that (1) all the parameters assayed are linked in that they all obey the same threshold value originally determined at the behavioural level, and (2) a unique and sharp transition temperature of 15 °C has been determined above which the various membrane changes connected with chemotaxis can be observed.

By way of illustration, let us summarise what might happen if a gradient of an attractant, say glucose, reaches the threshold of 0.1 mM at the plasmalemma of a plasmodium, which up to then was shuttling along at random (Fig. 44). Water structure tightens up, the outer membrane loses some negative charges and gets generally more pliable as the contraction force decreases; endoplasm gushes into the softened front, increasing motive force; hydrophobicity increases along with adhesion force, and positive taxis must result.

Much of this picture remains fuzzy, more so when we look at a repellent. The only clear differences are an increase, instead of a decrease, in the contractile force, and vice versa for the adhesion force. As a consequence, the signal at the membrane gets interpreted the other way around. If it can be shown that the outer membrane gets more rigid where the repellent strikes first, then the increased motive force might force pseudopodia out at the tail end and the whole creature moves away. This picture allows for many predictions that need to, and can be, tested. We should learn more about the important process of 'polarisation' in which transmembrane ion fluxes may be involved.

The main conclusion of the chemotaxis work is that there is no need to postulate specific receptors for the chemoreceptive membrane, a view that is shared from work on other protozoan and mammalian sensory organs (Seeman, 1972). Apart from chemoreception, similar results have been obtained for olfactory reception. Finally, in *Physarum* many of the specific parameters worked out above change accordingly in thermotaxis (Tso & Mansour, 1975) and phototaxis (Hato *et al.*, 1976). In the last case it was found that a plasmodium avoids the light. Of course it can only do that if the light is absorbed by a photoreceptor. As the avoidance of light was found to depend on the quality of the light, an action spectrum could be established (Bialczyk, 1979). Blue light is most active, indicating a blue-light receptor. The remarkable thing about that action spectrum is that quite comparable results have been obtained on the 'behavioural' level for *Physarum*, for fungi (*Phycomyces* or *Neurospora*), for a green plant (*Pisum*), and even a fly (*Drosophila*). In none of these examples of blue-light effects has the pigment been identified, hence it has been named 'cryptochrome'. However, while light has been implicated in triggering and maintaining some rhythmic processes in other species, mitotic synchrony and shuttle streaming in *Physarum* occur in the dark. However, in response to blue light the shuttle shows a decrease in tension force and may get out of phase (Block & Wohlfarth-Bottermann, 1981) and, as we shall see below, the same light receptor is clearly involved in a decision-making process in the sporulation pathway (Rakoczy, 1980). While motility and chemoreception seem to be

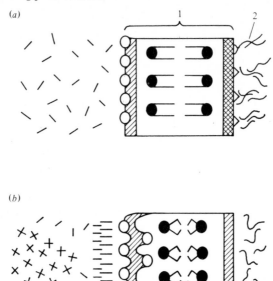

Fig. 44. Illustration of some events leading to positive chemotaxis in *Physarum*. (*a*) At the resting state, the surface (left) is exposed to liquid, the membrane (1) has a negative surface charge and the cortex (2) at the cytoplasmic face of the membrane contains a network of microfilaments, which help to shape the plasmodium (and prevent its rupture). (*b*) As an attractant (1) reaches threshold concentration, the water structure becomes more rigid (2), the surface undergoes a conformational change (3), some negative charges get buried (4) and hydrophobicity increases (5), all of which results in a decreased adhesion of the outer surface (6) and less tension generation inside the membrane (7). Consequently, a pseudopodium will be formed pointing towards the source of attractant.

controlled by an unspecific sensor mechanism, when two plasmodia accidentally come into direct contact the response can be quite specific. This leads us directly to the question of compatibility.

Incompatible plasmodia

As we know from the life cycle of *Physarum*, there are two kinds of cell fusion that can occur: one is the sexual fusion of amoebae (as gametes), the other is somatic fusion of plasmodia.

Sexual fusion may speed up the evolutionary process, by bringing together different individuals in a population to recombine their genes in a common gene pool. Consequently, after meiosis, zygote formation and subsequent sexual fusion, progeny will become different as outbreeding is promoted. On the other hand, somatic fusion of different individuals would also recombine two genomes in an individual (as a heterokaryon, or after a 'para-sexual' fusion of the two somatic nuclei). This does not happen in *Physarum*, however, as only plasmodia of identical genetic status fuse, and consequently no new genes become reshuffled. It can be argued, since fusion of like plasmodia (true siblings) is a natural process in development (though not an obligatory stage in the life cycle), that a fusion barrier prevents the horizontal distribution of an infectious disease (say a virus) over a large area inhabited by different races of one slime-mould species. Thus, somatic incompatibility could also aid the evolutionary process.

At any rate, although the reasons are not fully understood, it can be stated that in sexual development the one multi-allelic, complex mating-gene allows only gametes that differ in their gene make-up to develop properly, while a multigene system provides an insurmountable barrier for plasmodial fusion, unless two individuals are completely identical in the fusion-gene pattern. Here we have, of course, some formal similarities to the recognition of self, and rejection of foreign, in skin transplants of higher organisms. Fusion incompatibility has been aptly reviewed in *Physarum* by Carlile (1973, 1976), Carlile & Gooday (1978) and Collins (1979); the latter author also includes work on *Didymium*.

In former times, because so many plasmodia of different species look alike (and sometimes sporangia have been hard to obtain in the field), fusion incompatibility was used as a phenotypic marker to distinguish slime mould species one from another. However, since it was found that amoebae from two such plasmodia could mate successfully, they must have belonged to the same species. It seems that almost all specimens collected at different locations belong to different races, by these somatic fusion criteria.

Another important observation has been made: if plasmodia from these different geographical locations are put together, they ignore each other, and they can be maintained side by side on the same dish. If, however, spores from all these different strains are mixed up, one would expect that the whole spectrum of different plasmodia will arise, just as different as their parents were. That is not observed; only one strain survives, the rest being eliminated in a fierce battle in which incompatibility means more than non-fusion, it means plain killing.

In a well-controlled experiment with *Didymium*, a haploid amoeba of one mating type (say A_1) was crossed with a polyploid amoeba of a different mating type (A_2). Three kinds of plasmodia were obtained: haploid, diploid, and polyploid. When they produced sporangia from their F_1 generation, the haploid one led to haploid plasmodia of mating type A_1, the diploid also to haploid amoebae of mating type A_1, and the polyploid one to polyploid amoebae. The interesting case is the diploid plasmodium which yielded only haploid amoebae of one mating-type parent; obviously, the other had been eliminated. Consequently, one way of stabilising one genome over the other is chromosome elimination.

In this situation it could take a while for this process to become stabilised, and it could be that this is happening generally in *Physarum* plasmodia. It seems that the chromosome number is quite unstable as it fluctuates from 20 to 200 within a plasmodium, the 'normal' haploid set being 40. Meiosis might be the time at which those chromosomes that are not essential are eliminated. However, some chromosomes are under stringent control: in a diploid plasmodium only one pair of the mating-type allele (like A_1/A_2) is allowed to survive in a natural race.

The other means to maintain a specific slime-mould race is fusion incompatibility, which has been observed in variants derived from all strains cultured in the laboratory (*Physarum* strains Wisconsin, Indiana, and Colonia).

In *Physarum*, two sets of genes have been identified by the fusion assay. Each set contains several genes which are separate entities, and they are not linked to each other. These genes have been named *fus A*, *B*, and *C* ('fus' for fusion). Each gene exists as two alleles that are either dominant or recessive (for *fus B* and C), or co-dominant (for *fus* A_1/A_2). These gene functions prevent fusion, if any of them is not present in at least one dominant copy in both plasmodia.

Three other genes, called *let A*, *B*, and *C* ('let' for lethal reaction), do not seem to prevent fusion but cause a lethal reaction afterwards. Again, as long as one dominant copy is present in both partners, fusion yields intact plasmodia (i.e. AA, BB, $CC \times Aa$, Bb, Cc will survive). Yet, homozygous AA in one plasmodium and aa in the other leads to the lethal reaction. Plasmodium A will finally lyse plasmodium a. A bidirectional killing effect can follow, after plasmodia of the following constitution, $Ab \times aB$, have fused.

There are probably many more genes involved, and in the related species *Didymium* 13 different loci have already been determined. Again, they can also be classified as two groups: one affecting fusion, the other post-fusion reactions. However, in each group some genes have stronger effects than others. This leads to an elegant way to combine any mixture of these genes and predict the outcome of the incompatibility reaction (Clark & Collins, 1973; Ling, H. & Ling, M., 1974). This sort of genetic dissection has allowed four steps to be distinguished, since they get blocked consecutively. In the first, no fusion occurs; in the second, (as a fast and local effect) a clear zone appears within 30 s where fusion has occurred; in the third, a slow appearance and spreading of the clear zone is detected, beginning at 6 h postfusion; and as the last step, no effect is visualised until 24 h, when the nuclei of the loser have become eliminated (in another example of heterokaryon instability).

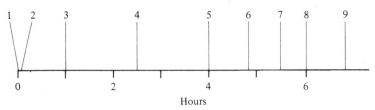

Fig. 45. Some events associated with the lethal reaction of a killer strain on a sensitive strain. 1, Surface contact; 2, plasmodial fusion; 3, commitment (point of no return, PNR); 4, end of transcription requirement; 5, end of translation requirement; 6, blockade of shuttle streaming; 7, inhibition of RNA- and protein synthesis; 8, onset of lysis; 9, breakdown of sensitive nuclei.

One way to interpret this set of observations is to distinguish discrete effects: no fusion, fast or slow post-fusion lysis, or heterokaryon incompatibility. Another interpretation of the same results – consistent with strong and weak gene sets – goes as follows: no fusion means fast reaction, when many genes are different between the two plasmodia. Then nuclear elimination is a very slow reaction, as only one of the many genes required for compatible fusion may be different. This interpretation assumes that the least problems arise from two individual plasmodia if they are as different as possible in the multiple set of genes that determines recognition of self. It is obvious that the first interpretation is quite different, as slow killing would seem to be more grave an interruption of life than no fusion (see Carlile, 1978 for further discussion).

In one case the sequence of events has been carefully analysed in *Physarum*, as killer strain 15 exterminates the sensitive strain 29 (Schrauwen, 1979). By separating the two incompatible plasmodia by a dialysis membrane or a millipore filter, it was shown that diffusable factors are not released from the killer to kill the partner; fusion and partial mixture of protoplasm seem to be required (Fig. 45). However, killer strain 15 has done its work after 1 h, seeing that strain 29 gets killed even if all material contributed by the killer is removed. Lysis of plasmodium 29 begins after 6 h and spreads over the whole plasmodium, and then the nuclei become degraded. Lysis does not require DNA synthesis, but it needs RNA transcription until 2 h after fusion, and protein synthesis until 4 h after fusion. At 5 h after fusion nucleic acid synthesis and protein synthesis begin to decline rapidly, shuttle streaming stops, and the surface becomes shiny. At 6 h lysis begins and spreads. This series of reactions occurs only if the plasmodia are well nourished. If the nutrients are withdrawn from the agar, none of the lethal reaction can be seen, but within 24 h the nuclei of strain 29 have become eliminated 'gently', so to speak.

Two more questions have been asked: are all nuclei eliminated in a lysed plasmodium, and in view of the elegant somatic fusion experiments joining mouse and man and plant cells by disrupting their cell membranes, how can incompatible plasmodia be fused anyway? With regard to the first question, viable plasmodia have been obtained in some cases from the clear area of a 'lysed' plasmodium (Ling & Upadhyaya, 1974). Interestingly, they contain nuclei which display a dominant gene towards the respective lethal effect gene of the killer strain. Such a particular gene must not have been expressed before the fusion, or a lethal reaction would not have taken place at all. Therefore, some dominant A-alleles must have been present among the millions of a-alleles in the many nuclei of the plasmodium which was at first unable to respond to the assault of the A gene effect of the killer strain. As to the other question, two plasmodia have been force-fused by gentle homogenisation of two batches of microplasmodia (say *fus AA* with big nuclei, and *fus aa* with small nuclei) which would never fuse on their own. After ten days some plasmodia could be isolated which contained both nuclear sizes. They could fuse with each other, but they did not fuse with either 'parent' (Jeffery & Rusch, 1974). This result, just as in other heterokaryon cells and cell hybrids, seems to indicate that, once the membrane barrier has been overcome, two different genomes and two different cytoplasms can get along fairly well.

It would be important to find out if such experiments work also with the killer strain. If these are really more closely related to each other than to those strains that just do not fuse with each other (as deduced from the genetic dissection above), such experiments (i.e. force fusion) should yield many viable heterokaryons between a killer and a sensitive strain.

One approach has been to put a porous membrane in between a killer and a sensitive plasmodium (Clark, 1980). Although the two plasmodia fuse with each other through the pores, which were too small to allow the transfer of nuclei, no lethal reaction occurred. Does that mean that the nuclei must carry the toxic message? One can speculate that the mixed membrane switches the vital intracellular coordination by some sort of 'membrane signalling' from killing to coexistence.

At any rate, compatibility studies in *Physarum* will help to find a possible mechanism of speciation for the taxonomists, and provide a definitive example of elimination of a predictable set of nuclei. This may warrant a search for specific restriction enzymes which have yet to be discovered in any eukaryotic cell, but have been postulated in other incompatibility systems of higher cells (Sager & Kitchin, 1975).

Finally, homogeneous material is available to probe the question of cell surface-mediated recognition of self in a combined genetic and biochemical approach. As in other recognition mechanisms in animal cells (cell aggregation, cell sorting, syncytium formation in skeletal muscle, histocompatibility and other immune reactions), the outer cover of the cell surface, including surface-bound lytic enzymes and the carbohydrate containing glycocalyx, must be integrated in the analysis, and we may yet discover a useful and telling function on the slime mould's slimy surface.

6

Changing the plasmodium

Although plasmodia have been kept growing for years by subculturing at regular intervals, a hungry plasmodium cannot survive unless it differentiates. If the nutrients are used up or, more generally, under any kind of unfavourable external condition, a plasmodium begins to cannibalise itself by degrading all its structures, including nuclei. However, at the same time it manages to construct two macroscopic forms for survival and propagation, either macrocysts or fruiting bodies. While the former can return directly to the plasmodial state, the latter can reach that state only after going through a microscopic unicellular organisation, the amoeba–flagellate state.

The microscopic stage can also become differentiated, either reversibly into a non-growing dormant cell, the microcyst, or irreversibly into the macroscopic plasmodium, which is at first a multinuclear structure until it breaks up again into multinucleated cells (the macrocysts) or uninucleated cells (the spores).

As stressed earlier, the alternative states of development can be induced in the laboratory and analysed on homogeneous material. We shall first discuss the macroscopic stages, in which many structural and metabolic changes have been observed that may be essential for differentiation, either following or preceding alteration in genome expression. Later on, a discussion of the microscopic stages will demonstrate the power of genetic dissection of differentiation processes. A recent and concise review of differentiation of *Physarum* has been provided by Gorman & Wilkins (1980).

Macrocyst formation: a reversible differentiation process

In its natural habitat a plasmodium moves over its substrate until any one of three conditions forces it to differentiate: lack of nutrients, temperature change, or desiccation. Differentiation begins at the tail end of the plasmodium where, after the shuttle streaming has ceased, portions of the protoplasm round up into little balls, first described as 'spherules' by Brandza (1926). They are held together by a network of fibres which is derived from massive slime production. The whole structure resembles a dry crust. It has been called a 'sclerotium', and can be stored for years, while the spherules within remain viable. These cysts are quite resistant and have been known to survive for ten weeks at 60 °C or three months in absolute ethanol.

The cysts can be brought back to life by adding water. A small plasmodium emerges from each spherule, shuttle streaming resumes, and the plasmodia fuse with one another, making plasmodial fusion a regular (though not obligatory) part of the life cycle of *Physarum*, thus leading to a syncytium from genetically identical units.

Many unphysiological adverse external conditions cause encystment. Among them are low pH (pH 2), heavy metal ions, or osmotic shock (e.g. 0.5 M sucrose) (Jump, 1954). It seems that any factor that can knock a plasmodium off-balance can trigger a monotonous differentiation and survival process. However, the morphological changes mentioned so far do not describe the differentiation process adequately, as can be best seen from the effect of various temperatures on a plasmodium (Hodapp, 1942). Plasmodia grow well at 22–28 °C, but they cannot live below 6 °C nor at 35 °C. If they are kept at 7–10 °C, it takes only 18 h for a sclerotium to form, and the fastest differentiation into spherules takes only 3.5 h, when a plasmodium is kept at 35 °C. Although these cysts look just like normal cysts that take a few days for development, they are not. While normal cysts can immediately be brought back to life after they have formed, the rapidly formed cysts will die unless they are allowed at least 24 h for maturation. This observation alone indicates that morphological changes are only part of the differentiation programme, and that controlled conditions are required to study encystment properly.

Most work has been done on microplasmodia grown in semi-defined medium in submerged culture. They are collected by gentle centrifugation and re-suspended in a non-nutrient balanced-salts medium (Daniel & Baldwin, 1964; Goodman, Sauer, Sauer & Rusch, 1969). The transformation of microplasmodia into cysts occurs with fairly good synchrony, within 36–48 h, and this process of 'spherulation' has been extensively analysed. Clearly, starvation is the trigger for differentiation under these conditions (Hüttermann, 1973). Some work has been done on macrosysts that have been induced from microplasmodia in fully defined growth medium, by adding 0.5 M mannitol (Chet & Rusch, 1969). It has been supposed that the starvation regimen triggers two series of events, differentiation and starvation, while the mannitol system only reveals one programme, namely macrocyst differentiation. Although such a clear-cut distinction has not been established, a number of similarities and differences between the two systems have been revealed, which suggest that differentiation does not proceed as a fixed and unique sequence of events. At any rate, macrocyst formation occurs in a closed system, and all building material and energy to construct the macrocyst must be derived from existing parts of the plasmodium. Furthermore, since germination and biosynthesis of macromolecules can be initiated during excystment in plain water, some storage of energy and precursors must also be provided during differentiation, just as in plant seeds or animal eggs.

What has been observed during macrocyst formation? The most general observation is that plasmodia immediately stop growing when they are suspended in salt medium (or induced by any other means to differentiate). This sounds like a trivial statement, but it is not, since plasmodia, if kept to themselves, can continue growing for several hours without nutrients, and reach the required protein:DNA ratio to undergo mitosis, just like well-nourished controls. Consequently, cessation of growth is initiated, despite the capacity to synthesise protein. It has been argued that a temporal dissociation of polysomes is an early reaction to the exogenous 'triggers' (Bernstam, 1978).

Several changes have been observed in the electron microscope (Goodman & Rusch, 1970; Zaar & Kleinig, 1975). The glycogen granules, which out-number ribosomes in the growing plasmodium, disappear over the first 12–15 h of starvation. By 18 h, the Golgi apparatus has become very active. At that time the production of slime (a galactose polymer containing sulphate and phosphate groups) increases; this polysaccharide is contained in Golgi vesicles which then fuse with the plasmalemma as they export the slime. This material holds much water and may cause a substantial dehydration of the cytoplasm and cessation of streaming of the endoplasm which probably turns into a gel. By 24 h, for the first time in *Physarum*'s lifetime, membranes of the endoplasmic reticulum (ER) can be seen within a plasmodium. At that time cleavage sets in, breaking up the plasmodium into many multinucleated units. This process has been formally viewed as a cell division occurring in the absence of nuclear divisions. This would complete the cell cycles that have been evident only as nuclear divisions during the growth phase. However, unlike animal cells during cytokinesis, the membranes do not form by cell constriction. Cellularisation in *Physarum* involves fusion of vacuoles, beginning in the middle of the plasmodium and moving outward. It has been clearly shown that most of the vacuoles, which are actually plasmalemma invaginations that permeate the whole plasmodium before cellularisation, disappear just as the cleavage membrane appears (Zaar & Kleinig, 1975). The invaginations vanish altogether as the surface membrane of each macrocyst rounds up, and the unit becomes pushed apart from its neighbours as more slime is secreted. During cleavage of the plasmodia, which lasts about 6 h, nuclear chromatin becomes less condensed, and by 30 h the ER becomes studded with ribosomes. By 36–48 h after suspending microplasmodia in salt medium, a two-layered wall has formed around each unit, and differentiation of macrocysts is complete. The main component of the cell wall is a polymer of galactosamine that contains phosphate, and about 7% of the wall by weight consists of protein. During this whole process a considerable amount of macromolecules disappear from the plasmodium: over 70% of the glycogen by 12–15 h; and by 32 h only about one half of the protein, one-third of the RNA, and two-thirds of the DNA remain. Surveying the morphological changes, we can assume that consider-

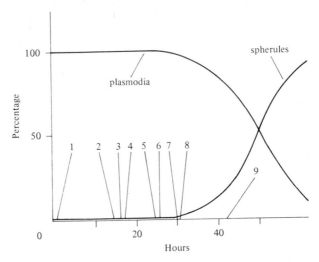

Fig. 46. Some events occurring during macrocyst formation (spherulation). 1, Halt of growth; 2, glycogen store depleted; 3, Golgi apparatus prominent; 4, increase of slime production; 5, appearance of endoplasmic reticulum; 6, cleavage (cellularisation); 7, rough endoplasmic reticulum; 8, dispersed chromatin; 9, wall formation.

able alterations must occur in the metabolism of phospholipids (as membrane components), of carbohydrates, as well as an increase in the turnover of protein and nucleic acids. Some of the events are sketched in Fig. 46.

The question remains: which of these general observations are significant for the differentiation process? It seems that most are required, at least to some degree, if we compare starvation with the mannitol system and employ reversible inhibition of macrocyst formation. All degradations occur if mannitol is added to growing plasmodia, albeit to a lesser degree. As differentiation sets in faster under these conditions, there seems not enough time to eliminate superfluous membrane and glycogen by degradation, and both structures get packaged in the vacuoles and become extruded as vesicles upon cleavage of the plasmodium.

A selective yet reversible inhibition of spherulation follows the addition of either ethanol (0.5%) or *n*-butanol (0.05%) to the medium (Kleinig, 1974). Under these conditions, the turnover of phospholipids does not occur, nor the net increase of lipids thereafter, and cleavage of the plasmodium is altogether blocked. We may speculate from this that the membrane is involved in sensing adverse environmental conditions to initiate differentiation but has become desensitised by the alcohol.

Another reversible inhibition of spherulation can be elegantly achieved simply by illuminating a starved suspension of microplasmodia. Light has been considered as an antagonist to growth in *Physarum*, inhibiting overall synthesis of macromolecules (Daniel, 1966). In the white mutant strain (Whi 1) four effects have been observed that are all mediated by the light, and as the action spectrum suggests, by the same blue-light receptor (possibly a flavoprotein) (Schreckenbach, Walckhoff & Verfuerth, 1981). At 450 nm, microplasmodia in non-nutrient salts medium take on a different shape; spikes and nodules appear on the surface. (These may be their 'tail ends', since the endoplasm gathers in the middle of the plasmodium hiding from the light, an abortive photophobic reaction.) It could also reflect a futile attempt to sporulate (yet another effect of the same photoreceptor, which we will come back to later). Those starving plasmodia do not differentiate. They remain viable for days, but finally lyse and die. In addition, they utilise glucose differently when illuminated.

Indeed glucose, if added to the medium, does not suppress a spherulation programme, once it has been triggered. This shows that differentiation requires more than starvation and darkness to become effective.

We begin to see that differentiation in *Physarum* is quite complex. It seems to require a signal, the membrane, a metabolic switch of glucose metabolism, degradation of polysomes, and an elimination of ribosomes. This, consequently reduces the capacity to make protein which in turn slows down the cell cycle. How about the genome as a possible master regulator of differentiation? The central hypothesis of differential gene expression predicts the need for selective RNA synthesis to explain any differentiation process. A simple experiment, involving actinomycin D, tells us that RNA synthesis may not be involved in macrocyst formation at all, but as this drug is not very efficient (even at $300\mu g/ml$, not more than 80% inhibition of uridine incorporation is achieved during starvation). On the other hand, experiments with cycloheximide demonstrate that spherule formation can be blocked as late as after 20 h in salt medium which is just prior to the morphological elaboration of the walls surrounding the spherules. This definitely shows a requirement for protein biosynthesis.

After these general remarks let us now consider the metabolism of carbohydrates and protein and thereafter the question of regulation of gene expression and some concepts of differentiation.

Carbohydrate metabolism

For the growing plasmodium, glucose is the most efficient carbon source. Maltose and soluble starch can also be utilised well. Galactose is a poor substrate, even though it is chemotactically active. Furthermore, amino acids are not converted into carbohydrate by gluconeogenesis as long as the mould is growing.

Glucose can be converted to glycogen, and from glycogen, glucose units can be released as glucose-1-phosphate by the key enzyme glycogen phosphorylase (Fig. 47a). This enzyme, which governs both the synthesis and utilisation of glucans, has many properties in common with other phosphorylases: it is a dimer consisting of two subunits, each of about 10^5 daltons, and has a similar amino-acid composition and immunological cross-reactions; but the regulation of its activity is quite unusual. While in all other organisms this enzyme becomes phosphorylated, converting an inactive form into an active form, no such changes have been observed with the enzyme purified from *Physarum* (Nader & Becker, 1979). Instead, glucose-1-phosphate, the product of the enzyme reaction, is inhibitory, as is UDP-glucose which is derived from glucose-1-phosphate and UTP in a reaction involving pyrophosphorylase. Inhibition by glucose-1-phosphate is effective in both metabolic directions: synthesis of glycogen and utilisation via glycolysis. The only other organism known to have such enzyme characteristics is *Dictyostelium*, where carbohydrate metabolism is also involved in the differentiation programme (Wright, 1973). Nevertheless, in *Physarum* inhibition of glycolysis by glucose-1-phosphate can be counteracted by AMP (at least *in vitro*), and the pathway of glycolysis is open in the growing plasmodium where oxygen is consumed and CO_2 is produced from glucose. The pentose phosphate pathway is also operative in *Physarum*, but not as intensely as glycolysis. UDP-glucose can be polymerised to glycogen by glycogen synthetase, or it can be converted by epimerase to UDP-galactose. This is the precursor for the synthesis of slime (see Fig. 47 for some inter-relations of these reactions).

This is the general situation during growth. What happens after the plasmodium is put in salt medium and gets hungry? First of all, a rapid drop in glycogen occurs. This is achieved without an activation by phosphorylation of the phosphorylase, and no new enzyme is synthesised either, as has been checked by titration of the enzyme protein with an antibody. The decrease in glycogen is paralleled by a decrease in glycogen synthetase. The glucose derived from that store does not appear in the slime which begins to be accumulated only after most of the glycogen has already disappeared (Fig. 47b).

During starvation oxygen consumption is reduced by one half, the pentose phosphate pathway is favoured over glycolysis and, if [^{14}C]glucose is added (remember this does not interfere with spherulation), no $^{14}CO_2$ is produced. Yet three different products become labelled, even in a nitrogen atmosphere: one is water-soluble and has been identified as trehalose; another is insoluble in water and is an unidentified glucan containing mainly glucose; and an acid is produced and secreted into the medium (Schreckenbach *et al.*, 1981). It is quite certain that neither glycogen nor exogenous glucose is a precursor of the abundant slime nor of the cell walls. Also, none of the enzymes mentioned in the pathway of UDP-galactose show a close corre-

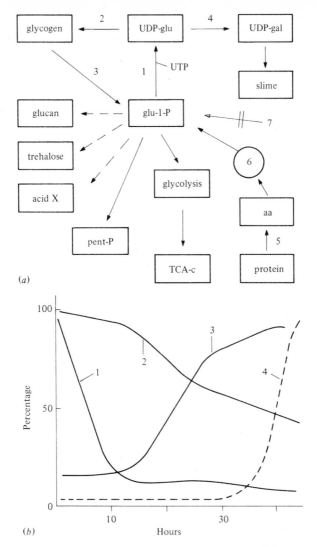

(a)

(b)

Fig. 47. Some metabolic changes during macrocyst formation. (a) Various pathways of glucose: glucose-1-phosphate (glu-1-P) can be utilised in glycolysis, thus feeding the tricarboxylic acid (TCA) cycle, or in the pentose-phosphate pathway (thin arrows). Glucose is also the source of trehalose, a glucan and an extracellular organic acid, particularly during starvation in the dark (dashed arrows). During growth glucose ends up in glycogen, involving pyrophosphorylase (1) and glycogen synthetase (2), from which it can be quickly released at the initiation of starvation by glycogen phosphorylase (3). UTP-glucose is also utilised in slime production, with UDP-galactose (produced by epimerase) (4) as one intermediate. However, the source of glucose for the massive slime production stems from protein degradation via proteases (5) and gluconeogenesis (6), even if glucose is exogenously applied (7) during starvation (broken arrow). (b) Temporal correlations of: glycogen utilisation (1), protein degradation (2), slime production (3) and spherule formation (4).

lation in activity to the increase in slime production. The key enzyme UDP-pyrophosphorylase even declines before slime production increases.

A simple calculation confirms that other sources must be tapped during spherulation (Hüttermann, 1973): 7 mg carbohydrate (including 5 mg slime) and 3 mg wall material (mostly galactosamine) are produced in a typical culture which contains only 8 mg of glycogen to begin with. However, at the same time 15 mg protein have been degraded, presumably down to amino acids. These must be utilised for de-novo protein synthesis, but also for gluconeogenesis. In this way, amino-acid derivatives could end up in the slime. This has been shown by labelling slime with exogeneous [^{14}C]glutamic acid, and inhibitors or stimulators of gluconeogenesis have either decreased or increased slime production. It has also been demonstrated, utilising labelled slime, that it could not be metabolised by a starving plasmodium (nor by any other organism), and that it is not a precursor of cell wall polymer of galactosamine.

An elegant experiment has shown that glycogen is not required at all for the elaboration of the spherulation programme (A. Hüttermann, personal communication). Plasmodia were grown in ethanol as carbon source and no glycogen accumulated. However, normal spherules, and lots of slime and wall material were made once the plasmodia were put in salt medium. The easiest assumption is that amino acids become available for gluconeogenesis as protein is degraded. Amino acid pools have been analysed in *Physarum* during spherulation. Although there were differences between various amino acid species, an overall reduction of the pools by two-thirds occurs in the first 6 h, and is followed by an increase which is caused by protein degradation, and another decrease as slime and wall material are synthesized. During differentiation, degradation of protein seems to alter the pool sizes of amino acids, and gluconeogenesis is the main pathway for differentiation-oriented metabolism. Indeed, gluconeogenesis operates only during starvation and cannot be induced during growth.

This view applies for *Physarum polycephalum*, yet things seem to be different in the very similar plasmodium of *Physarum flavicomum* (Henney & Chu, 1977). In that species, differentiation is also correlated with an imbalance of amino acids, which in turn is brought about by protein degradation and quick re-utilisation of the amino acids in energy metabolism and polysaccharide synthesis. However, while *P. polycephalum* is unaffected by exogenous amino acids, the addition of amino-acid mixture (or leucine in particular) to *P. flavicomum* blocks the spherulation programme. As a consequence, further degradation of protein is inhibited in *P. flavicomum*, but not in *P. polycephalum*. While the role of protein degradation in differentiation is evident for both species of *Physarum*, the exact mechanism of amino-acid control over differentiation remains to be established.

Before we turn to protein metabolism, let us recall that degradation of glycogen and accumulation of slime do not require protein synthesis, as shown with the cycloheximide experiments. These metabolic changes may not be part of the differentiation programme at all but reflect the state of hunger in a plasmodium. On the other hand, cycloheximide inhibits differentiation proper, i.e. the formation of macrocysts.

Protein metabolism

Proteolysis can play an important role in specifically altering proteins after they have been synthesised. Some enzymes are activated by the cleaving of pro-enzymes, and some proteins have to be tailored before they can be utilised in morphogenesis. As protein degradation is an irreversible modification of the protein pattern inside the cell, or at its surface, these changes may well break up regulatory circuits of vegetative growth and trigger new developmental pathways. The result could well be a differentiation process.

At any rate, hydrolysis of protein yields amino acids which can be used in various ways as we have discussed, including the synthesis of new proteins. This mechanism can become regulatory if the concentration of only one essential amino acid (methionine in *Physarum*) becomes limiting and needs to be supplied via degradation.

In *Physarum*, it has been estimated from isotope dilution experiments (Sauer, Babcock & Rusch, 1970; Wendelberger-Schieweg & Hüttermann, 1978) that after 24 h in salt solution, at the time when spherules appear, at least half of the protein present in the growing plasmodia has been degraded. Of the remaining protein about one-half has been newly synthesised. By similar methods it has also been shown that protein labelled during the growth state is more rapidly degraded than protein labelled during the differentiation phase. An unexpected result has been obtained with actinomycin D: while no inhibitory effect on amino acid incorporation was observed over a period of several hours during growth, a stimulation of protein synthesis was caused by this RNA synthesis inhibitor during the first hours in salt medium (Sauer, Babcock & Rusch, 1970). This result could be interpreted if RNA synthesis was required for the expression of a protease which causes the protein degradation in the untreated starving plasmodia. Proteases have been analysed in crude preparations, and in one study 12 enzyme fractions were detected which could be classified into three groups of iso-enzymes (Polanshek *et al.*, 1978). Little change occurred during spherulation, except for the enzymes in group one. While three fractions decreased over the first four hours in salt medium, one showed a marked increase in activity immediately after transferring plasmodia into salt medium (Fig. 48). It is not clear whether this protease activity is the result of an interconversion of some of the disappearing fractions or to de-novo synthesis. In a different analysis employing other substrates and different

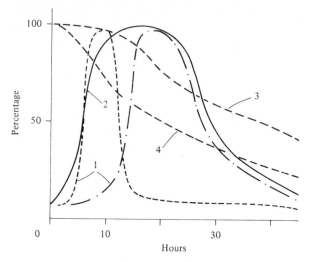

Fig. 48. Some correlations between degrading enzymes and degradation of macro-molecules. Two bursts of protease (1), increase in nucleases (2), amount of total protein (3) or total RNA (4).

parameters for in-vitro assays, ten putative iso-enzymes of aminopeptidases were detected, four of them only in spherulation (Hoffmann & Hüttermann, 1975). Employing density labelling with heavy amino acids, complete de-novo synthesis for these four iso-enzymes has been demonstrated, which begins after about 12 h in salt medium. It is not clear whether RNA synthesis is required for these enzyme activities to appear. Until the intracellular location and the respective substrates of *Physarum* are known, we can only state that various proteases, some old and some new, are present in the plasmodium, and could be responsible for the marked protein turnover. At the same time, we have obtained a good example of de-novo synthesis of an enzyme during spherulation, where net degradation prevails.

A similar picture emerges for the nucleases which might be involved in the substantial breakdown of RNA and DNA (Fig. 48). The two well-charac-terised extracellular RNases (Phy I and II) do not seem to change much dur-ing spherulation. In crude preparations as many as 12 intracellular RNases have been detected, and some fractions increase considerably during dif-ferentiation (Brand, Hüttermann & Haugli, 1975; Chet, Retig & Henis, 1973). The activity of single-strand- and double-strand-specific DNases have been observed both in growth and in starvation. While all these enzymes are active in acid milieu, an alkaline nuclease has also been detected, a zinc-metalloenzyme that rapidly increases during the first 6 h in salt medium and correlates well with the concomitant decreases in RNA. This increase is abolished by cycloheximide and may indicate another de-novo enzyme synthesis (Waterborg & Kuyper, 1979). Two examples of another enzyme, phosphodiesterase (PDE), which could be involved in nucleic acid degradation, illustrate two different patterns of protein regulation during differentiation. PDE I increases 18-fold early in the process, and density labelling indicates that this change in activity is due to a higher rate of synthesis than degradation (glutamate dehydrogenase behaves in a similar way). On the other hand, PDE II completely disappears during spherulation. Such behaviour may also be observed for the two enzymes already encountered in the mitotic cycle which are involved in polyamine synthesis (ornithine decarboxylase, and *S*-adenosine-L-methionine decarboxylase) (Mitchell & Rusch, 1973).

Yet, another example of an enzyme activity that changes during spherulation is UDP-glucose epimerase which supplies UDP-galactose units for slime synthesis: its activity level decreases (either its rate of synthesis declines or its rate of degradation increases), resulting in a lower level, paradoxically just when much slime is made.

To sum up, in the few cases where changes in enzyme activity are associated with differentiation, a complex picture emerges. While at least some new syntheses seem to be responsible for overall degradation processes, no definite protein synthesis can be attributed to distinct events in spherulation, thus there is no sign of 'luxury molecules' as yet.

Furthermore, the regulation of either synthesis or degradation, which leads to the enzyme patterns described, is not coordinately controlled by a supposed differentiation programme. This has been shown by interrupting spherulation after 12 h and placing the plasmodia back into complete growth medium: some enzymes do and some do not respond to the shift-up experiment (Hüttermann, 1973).

All these results are in marked contrast to those obtained for the differentiation of the cellular slime mould *Dictyostelium*. Here, definitive sequences for many enzymes have been documented, including several for carbohydrate metabolism as the cellulose cover of the pseudoplasmodium is formed (Loomis, 1975). Such a developmental programme, which has been shown to be regulated by selective gene expression, does not appear to exist in *Physarum*.

In *Physarum*, there is one way to ask whether the changes in enzyme activity detected in salt medium have anything to do with the differentiation programme at all. It becomes obvious, by comparing enzyme patterns in either mannitol or starvation medium that, although some of the changes in protease and nuclease seem relevant to differentiation, all other cases of increased enzyme activities are diminished in the mannitol system. As an example, PDE I and glutamate dehydrogenase, which both increase over ten-fold in salt medium, do not increase more than two-fold in the mannitol system, but macrocysts are made nevertheless.

In yet another test, mannitol added to plasmodia in salt medium was shown to inhibit differentiation, while the plasmodia starve to death. Under these conditions PDE I and glutamate dehydrogenase increase markedly, which clearly demonstrates that these changes in enzyme activity are associated with starvation rather than differentiation (Sauer, 1973 for review).

While practically all drastic changes in enzyme levels seem unrelated to the production of differentiation-specific structures, it must be stressed that about 50 more enzymes have been tested, and most either stay unchanged or increase by a factor of less than two during spherulation. This indicates that most genes assayed up to now are continuously expressed, just as in the mitotic cycle during growth. Furthermore, where an increase of a factor of two is observed, while overall protein turnover has eliminated half of the protein content, such increase in enzyme activity reflects merely its relative stability over other proteins (and no increase in its synthesis at all). Nevertheless, since cycloheximide interrupts the differentiation sequence, it seems that the relevant genes have not been distinguished as yet.

Before we discuss RNA metabolism, we should mention that some alteration in the composition of nuclear protein may have implications for the regulation of gene expression (as reviewed by Walker *et al.*, 1980). Of the nuclear proteins, although all histones are present, it has been estimated that during starvation more protein is found in band 4 of the gel-electrophoresis pattern. It contains histone H_3, which is one of the conservative histones and a component of the nucleosome core.

Significant changes do occur in a fraction of the non-histone proteins (NHP) during starvation. Although these proteins are abundant (about 10^5–10^6 per nucleus), the coordinated disappearance during starvation and reappearance upon refeeding seems to indicate that they may regulate overall chromatin structure and possibly its function. Generally, the NHP (which amounts to 4:1 in relation to DNA) decreases during starvation. In particular, three fractions are lost from the nucleolus. In addition, one protein, the species of 34 000 daltons, which is associated with nascent non-ribosomal RNA (as discussed in chapter 4), disappears completely during starvation. This must have consequences for RNA processing.

In a different analysis, the nucleolar minichromosome has been analysed during growth and starvation. The chromatin was fractionated into pieces which contain the ribosomal gene (Johnson *et al.*, 1979). As discussed for the growing mould, the fragment containing the rDNA gene is cut differently by micrococcal nuclease, yielding a slower sedimenting particle (peak A) which contains more NHP of the high mobility group (HMG). However, if chromatin has been derived from starving plasmodia, no peak A can be resolved. These results indicate that the ribosomal genes are openly transcribed in actively growing moulds but closed in a starving plasmodium. This leads us directly to the discussion of RNA metabolism.

RNA metabolism

More than 90% of the stable RNA in a plasmodium is rRNA; the increased degradation of RNA by nucleases during starvation described above mainly effects the disappearance of ribosomes. As shown for protein, RNA molecules that have been pre-labelled during growth are more rapidly degraded than those labelled during starvation. Furthermore, overall synthesis of RNA is less intense during spherulation as compared to growth. When the labile RNA is fractionated into $poly(A)^+RNA$ and $poly(A)^-RNA$, or separated according to size, it becomes evident that rRNA and hnRNA (including mRNA) are affected alike. Among the non-ribosomal RNAs, a higher proportion of smaller molecules has been detected during starvation. With the increase in nuclease activity and the expansion of endogenous nucleotide pools during RNA degradation, it cannot be calculated how severely the rate of synthesis is reduced. However, although all classes of RNA are affected, it is also clear that all RNA classes are synthesised. Most of the relevant data have been reviewed by Seebeck & Braun (1980), Melera (1980) and Turnock (1980).

In addition to these alterations, significant changes in the amount of a polyphosphate fraction have been observed. From a low level after 2 h in salt medium, an increase has occurred after 16 h, and the highest level (30% of all phosphorus compounds as polyphosphate) is detected in spherules. This material, which shows an inverse relationship to overall RNA synthesis (Goodman *et al.*, 1969), can be looked on as a possible energy and phosphate storage product since upon refeeding a fraction of this phosphate has been detected in newly made RNA.

Some insight into the possible regulation of RNA transcription has come from experiments *in vitro*, and it seems that the two main classes of RNA, rRNA and hnRNA, are controlled quite differently by their respective RNA polymerases. At first, after solubilisation, RNA polymerase A shows a marked decrease and is practically absent half-way through the differentiation period, at a time when the level of the soluble RNA polymerase B has remained unchanged (Hildebrandt & Sauer, 1976c). Later on, in the completed spherule, polymerase A activity is once again present, as is polymerase B.

The decrease of RNA polymerase A activity parallels the increase in the nucleolar initiation inhibitor, described above. This inhibitor accumulates over ten-fold during starvation, up to 10 $\mu g/10^7$ nuclei, which is enough to completely inhibit *in vitro* all RNA polymerase A extractable from a plasmodium with that many nuclei. This inhibitor might act *in vivo*, since an inactive RNA polymerase A prepared from plasmodia after 20 h in salt medium, can be reactivated *in vitro*, presumably after release of the inhibitor (Hildebrandt & Sauer, 1977a). However, these results do not explain why inhibition of rRNA transcription continues in spherules, when the inhibitor is no longer present (Hildebrandt & Sauer, 1976b).

Whereas the soluble RNA polymerases behave as continuous enzymes, as do most proteins analysed during starvation, the endogenous RNA polymerase B activity of isolated nuclei decreases at the same time as incorporation *in vivo* into non-ribosomal RNA becomes reduced. There is some evidence for two regulatory protein factors, one positive and one negative control element, which could be involved in the inhibition of over 80% of nuclear RNA polymerase B activity: the elongation factor disappears during starvation (Ernst & Sauer, 1977), and a small non-histone protein appears (Hildebrandt & Sauer, 1977b). It may be significant that this protein fraction inactivates RNA polymerase B only in nuclei from a growing plasmodium, but cannot further inhibit the residual activity of that enzyme tested in nuclei isolated from starved plasmodia. Finally, there is some circumstantial evidence that the large subunit of RNA polymerase B may become cleaved by proteolysis from 210 000 to 170 000 daltons in starved plasmodia (Smith & Braun, 1978).

All information on RNA discussed up to now pertains to quantitative aspects, but qualitative changes have been looked for by various methods. Originally, RNA–DNA-hybridisation, under conditions where only repeated DNA sequences reacted, did not reveal any differences if spherulation occurred in salt medium. However, in the mannitol system, evidence for a decrease of some and an increase in other RNA sequences has been obtained (Chet, 1973). In a different approach, poly(A)$^+$RNA from starved or growing cultures has been hybridized in high concentration to isolated single-copy DNA, which has been highly labelled enzymatically *in vitro* by nick-translation. Under these conditions it has been shown that the RNA composition during starvation has become less complex: up to one-half of the sequences seem to be absent during starvation. These findings are consistent with the determination of the sequence complexity of poly(A)$^+$RNA by hybridisation to cDNA. In the homologous reaction, where both these components were derived from growing or starving plasmodia, respectively, poly(A)$^+$RNA from the latter cultures contained about half as many different nucleotide sequences as compared with the growing state. In the heterologous reaction, where cDNA has been obtained from a growing plasmodium and poly(A)$^+$RNA from a starved plasmodium, about one-third of the RNA sequences present in the growing culture have not found a complementary sequence in the RNA from the starved culture. Furthermore, the sequences common to both stages seem to become diluted during starvation (Baeckmann, 1980). That result may explain the rate of incorporation of amino acids into protein in the presence of actinomycin D (Sauer *et al.*, 1970). While no effect of the drug can be observed over at least 6 h during growth, after 20 h in salt medium the addition of actinomycin D drastically reduces the amino-acid incorporation (by over 50% within 4 h in salt medium, as well as in the mannitol system). During that time of development the attachment of ribosomes to the endoplasmic reticulum occurs, as seen in the electron microscope (Goodman & Rusch, 1970), a process that is also inhibited by actinomycin D. However, it should be stressed that seemingly normal spherules are made in the presence of the drug, and preliminary experiments of in-vitro translation into polypeptides followed by SDS gel electrophoresis have not revealed any new proteins during spherulation, nor have any disappeared (Baeckmann, 1980).

Although analyses of complex RNA mixtures are not very accurate, these results indicate that, if anything, the RNA becomes less complex during differentiation. In general terms, spherulation may be regulated by inactivation of the genome rather than by its activation. However, we cannot exclude that a few genes have to be activated to ensure the differentiation process. Only after appropriate mutants become available, can this question be followed up.

Nevertheless, we can conclude that gene expression is required during the differentiation process in *Physarum*, in a similar way to oogenesis in higher animals (Davidson, 1976). One finding is that actinomycin D, although it does not block the formation of walled cysts, prevents the reversal of the encystment state: it yields non-viable plasmodia after excystment. On the other hand, plasmodia that excyst from normal macrocysts do so even in the presence of actinomycin D. They seem to contain a store of mRNA which can be utilised after excystment, which can occur in plain water and does not require any exogenous nutrients for several hours; yet there is net synthesis of RNA and protein, and DNA is synthesised directly upon germination.

However, it should be noted that these findings, in *Physarum* as in eggs, do not imply that the RNA species, which have accumulated at one time to be utilised later on (after years), are any different from those present in other stages of development.

We have just mentioned that DNA synthesis is among the early events that occur after excystment which raises a new question for the analysis of macrocyst formation as a typical differentiation process: what is the state of the mitotic cycle?

The state of the mitotic cycle

Spherules that arise in salt medium, or in the defined growth medium in the presence of mannitol, readily excyst if fresh medium or just water is added. Over 90% of the spherules excyst in a process which takes about two to three days. This is also the case if any unfavourable condition has triggered macrocyst formation, unless the whole process has taken less than 24 h. If it takes less time, e.g. at high temperature, a maturation period of one to two days is required before viable plasmodia can be obtained from spherules.

The uptake of water builds up osmotic pressure, the wall ruptures, and the plasmodium, awakened from dormancy, starts shuttling and moves away, leaving the wall behind. This is a very dangerous decision for the plasmodium to make, because it cannot know whether nutrients will be available. If there are none, it will be next to impossible to make a new wall quickly and encyst once more. In the laboratory a 0.1 M sucrose solution prevents excystment, even in the presence of other nutrients. However, in nature the control over dormancy may be brought about by a negative control element, the germination inhibitor, which has been detected in the medium in which macrocysts have formed. This material has been described as a heat-resistant peptidoglucan of low molecular weight, which has not yet been purified (Chet & Hüttermann, 1977). In this context it should be mentioned that among the extracellular polysaccharides, slime comprises less than half (Hüttermann, 1973). Among the proteins, two renin-like proteases have been detected and the two RNases already mentioned.

In addition, it has been observed that bacteria do not grow if extracellular slime is present. Even more interesting is the fact that *Physarum* amoebae are inhibited by the same material in their cell division (Henney & Asgari, 1975). The active principle, which has been enriched in *Physarum flavicomum*, is heat-sensitive and inactivated by papain (a protease). The blockade of cytokinesis occurs at very low concentration. Synthesis of RNA, protein, and DNA are inhibited as well, while respiration increases in the presence of this inhibitor. Formally, this situation is similar to that in tissue culture cells, or tissues in an animal, where the depletion of serum factors or the level of self-regulating factors (such as tissue-specific chalones) seem to control the cell proliferation.

We can now ask whether the inhibition of growth associated with macrocyst formation and the re-initiation of growth during germination of cysts show similarities to the very important developmental controls required for coordinated cell proliferation in a multicellular organism. Again in formal terms, the cells in all tissue culture systems can be said to accumulate after mitosis and before initiation of S phase when proliferation is inhibited. They become arrested in G_1 phase (or, if it lasts very long, in G_0 phase). A large number of events must happen in response to refeeding or addition of growth factors (including uptake of small molecules and synthesis of various macromolecules) until DNA synthesis occurs. This defines the transition from G_1 to S phase, which is regularly followed by mitosis. Similarly, lymphocytes have become a model system. They are quiescent cells which can be stimulated to proliferate unspecifically in a Petri dish by plant lectins, such as phytohaemagglutinin. In the human body, the B-lymphocytes are specifically stimulated by antigen, and they proliferate before they produce antibodies.

In my view, the relevant observation in *Physarum*, with respect to proliferation control, comes from the few studies on DNA synthesis during starvation and resumption of growth after excystment. A mitosis has been described at about 5 h of starvation, which correlates with a peak of thymidine incorporation into DNA during 4–8 h in salt medium. Thereafter, DNA synthesis declines and DNA content, after 24 h in salt medium, has decreased to almost half of the value measured at the beginning of starvation (Sauer *et al.*, 1970), on a per-culture basis. It is not clear whether this reflects degradation of those nuclei that may not have taken part in the only S phase detected in salt medium, or those that might become susceptible to the coordinated increase of the alkaline nuclease at 6 h in salt medium. It could also be explained if all nuclei enter a G_1 phase during spherulation. Once mature spherules have formed, a halving of both the nuclear diameter and the typical G_2-phase DNA content have been measured (DNA content: 0.6 pg as compared with 1.0–1.1 pg/nucleus) (Mohberg, 1974). If these nuclei are really in a 'G_1 phase', DNA synthesis must occur before the first nuclear division after germination takes place. Indeed, DNA replication has been observed before mitosis occurs with a fair degree of synchrony at 6 h after excystment (Fig. 49*a*).

The fluctuation of one particular enzyme, ornithine decarboxylase (ODC), a key enzyme in the synthesis of polyamines – substances which have been implicated in proliferation control – correlates well with the decrease of DNA synthesis during starvation and DNA replication after initiation of growth (Mitchell & Rusch, 1973). It is not yet clear whether the re-appearance of the enzyme takes place before or at DNA replication. This may be an important distinction in view of the hypothesis of replication–transcription coupling, operating during the growth phase of the mitotic cycle where ODC is synthesised early in S phase. Another telling enzyme that has decreased during starvation, thymidine kinase (TK), does not increase before DNA replication after excystment. This reinforces the findings described for the mitotic cycle: this enzyme, although a marker of a late G_2-phase event, is not essential for DNA synthesis at the putative G_1–S transition in excystment.

The hypothesis that nuclei in mature *Physarum* macrocysts are arrested in 'G_1 phase' would explain why a maturation period is required. One mitosis must be postulated to take place which, in contrast to the growth cell cycle, is not coupled to an S phase any more.

All work on macrocyst formation in *Physarum* has been done by putting microplasmodia under stress, which is again quite similar to tissue culture work. And, as in the animal systems, the experiments are done on heterogeneous material. Although the issues of spherulation and the cell cycle state are not settled in *Physarum*, they might be as sketched in Fig. 49.

However, recently a completely synchronous formation of macrocysts has

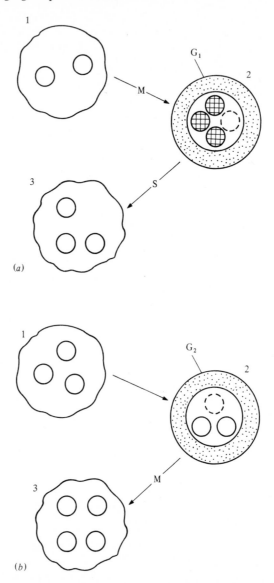

Fig. 49. Putative states of the cell cycle during spherulation. (*a*) A 'critical' mitosis occurs in the starved plasmodium (1) before mature cysts arise (2), which are arrested in G_1 phase (some nuclei degrade); upon germination (3) an S phase must occur before the first nuclear division. (*b*) A plasmodium (1) becomes transformed into a cyst (2) and then arrested in G_2 phase (some nuclei are degraded); upon germination of the plasmodium (3) mitosis occurs, followed by S phase.

been obtained by starving one well-fed macroplasmodium that has been in the process of its naturally synchronous cell cycle for 24 hours. This procedure, which involves the addition of 0.5 M mannitol to a macroplasmodium placed in salt medium, will now allow analysis of the process of differentiation and resumption of growth after germination. Now this developmental sequence can be induced at will and at defined points of the nuclear division cycle (Jalouzot & Toublan, 1981). This unique situation will also clarify whether macrocyst formation is really a simple 'reversible' differentiation process, not affecting the cell cycle, or whether the sequential events involve an irreversible step, since a G_1 phase cell cannot revert to mitosis unless a round of DNA replication takes place.

While these questions can now be tackled, we are left with a major problem. In *Physarum*, as in animal cell systems, it is not known how the many different signals of 'unfavourable' or 'favourable' environmental conditions lead to arrest or proliferation (Fig. 50*a*). In lower organisms, e.g. in sporulation of a bacillus, a transient increase in highly phosphorylated nucleotides has been observed (like guanosine-pentaphosphate) which, in turn, might selectively inhibit rRNA transcription. In higher cells a 'pleiotypic response' is evoked by any of the effective exogenous triggers which might act on the cell surface and mediate the signal somehow all the way to the cell nucleus, again reducing rRNA transcription and protein biosynthesis capacity. In addition, another unusual phosphorylated material, diadenosine $5',5'''-P^1-P^4$-tetraphosphate (AP_4A), has been identified which increases quickly by over 100-fold if cells get hungry (Grummt *et al.*, 1979), at least in tissue culture systems. This factor has been implicated in regulating the activity of DNA polymerase α, which even if present in a form that catalyses DNA synthesis on an exogenous template, cannot initiate replication on intact chromatin inside the nucleus. Preliminary experiments have revealed a 30-fold increase of AP_4A in *Physarum* following mitosis.

It has also been generally asked, in particular for *Dictyostelium* (by Wright, 1973) and for *Physarum* (Daniel, 1966), whether the exogenous 'trigger' changes first the metabolic state of a cell, which in turn affects the expression of the genome. That is, regulation would occur far away from the nucleus; for example, by substrate fluctuations, enzyme modifications, pH changes. It has also been asked whether the signal is first relayed to the information centre of the cell, the nucleus, and evokes a developmental programme which involves first gene expression (i.e. selective transcription, de-novo synthesis of regulatory proteins and enzymes) and secondarily affects the metabolic state. Which of these alternative views is correct has still not been resolved (Fig. 50*b,c*).

Furthermore, in *Physarum*, the other general view, that many signals are sensed at a single site and trigger a monotonous developmental sequence, can also be challenged, if we remember the various early effects that are

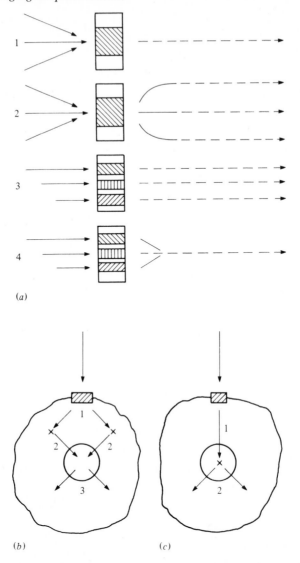

Fig. 50. Alternative pathways of spherulation. (*a*) Changes in the environment (arrows) are perceived by the plasmodium (probably at the outer membrane, hatched area), followed by intracellular events (dashed arrows) leading to differentiation. 1. Many signals, one sensor, one pathway. 2. Many signals, one sensor, many pathways. 3. Many signals, many sensors, many pathways. 4. Many signals, many sensors, one pathway. (*b*) Epigenetic control. The exogenous signal (arrow) may affect first a cytoplasmic target (1) which changes nuclear activities (2), which, in turn, affects the cytoplasm (3). (*c*) Genomic control. The exogenous signal directly changes the nuclear activities (1) which affect the cytoplasm (2).

brought about by an unfavourable environment before any morphological changes become apparent. The membrane might be generally affected, as ethanol, and more so *n*-butanol, prevents cyst formation. Alternatively, the membrane could be specifically affected, as a 'receptor' for mannitol has also been discussed. The polysomes in the cytoplasm have been implicated as a sensing mechanism, as well as the rRNA synthesis in the minichromosome of the nucleolus. A shift in glucose metabolism, protein modification and degradation, changing pools of amino acids, and a switch to gluconeogenesis have also been noted. Therefore, it could well be that many independent sequences take place which have been coupled to look like one definite developmental pathway (see Fig. 50*a*).

The observations outlined in this section have been made at different places under different conditions, and a coherent view has not emerged. However, one must now attempt to determine the interactions between the complex structural and chemical changes; mutants must be obtained and biochemically analysed in order to discriminate the general metabolic events from the controls of growth and proliferation, and, it is hoped, detect developmentally regulated reactions which could lead to an understnding of both differentiation and growth. To date, we must note that understanding macrocyst formation in *Physarum*, although a process dissociable from growth unlike most other models of development, has been hampered by the effects of starvation, many of which are unrelated to the developmental process. However, sporulation, the other survival alternative for a hungry plasmodium, may reveal a more clear picture, as it is only triggered after a long time of starvation, when the plasmodium is almost dead. Hence, it can be anticipated that anything it does in response to the developmental trigger must be relevant to our question, because a plasmodium that has become committed to sporulate has no means of reverting directly to the growth state: either it survives by differentiating, or it is destined to die.

Sporulation: an irreversible differentiation process

The sporangia of *Physarum* look very different from the growing plasmodium or the encysted sclerotium: they are the most obvious case of differentiation. A full account of the morphological changes has been given by Guttes *et al.* (1961), and metabolic changes have been reviewed by Daniel (1966). A comprehensive review has been compiled by Gorman & Wilkins (1980).

Nobody has been able to convert a fruiting body back into a plasmodium directly. Hence, this transition of the plasmodium is a uni-directional developmental process, which must be determined at some point along this pathway which covers an essential section of the life cycle.

Nor has anybody been able to convert a growing plasmodium into sporangia and, formally, one can say that the sporulation programme is repressed during growth. Hence, an extensive period of starvation is required to condition a plasmodium to get ready for sporulation, i.e. to become 'competent'. A plasmodium that has become competent to sporulate still has the option of forming a sclerotium (a crust with macrocysts) in the dark, or fruiting bodies in the light. We have already described that illumination prevents the differentiation of macrocysts; therefore, the two forms of differentiation may be mutually exclusive events.

On the other hand, while any starved plasmodium can form macrocysts in the dark, only a 'competent' starved plasmodium develops sporangia in the light. Consequently, starvation alone is not enough to condition a plasmodium. Therefore, a major problem, which at first sight seems to be remote from the morphological differentiation process, is to find out how *Physarum* becomes sensitised to react to light.

In its natural environment, a plasmodium, even a hungry one, avoids the light and prefers a moist dark place, which it finds under the decaying bark of a fallen tree, for example. However, a competent plasmodium searches actively for the light and settles down in a dry bright place.

In carefully controlled experiments with *Physarum nudum* it has been shown that the switch from negative to positive photoreaction is a dominant prerequisite, since a plasmodium will also sporulate in a moist atmosphere. It seems justified, from the action spectrum, to assume that the same photoreceptor (the cryptochrome) senses the light during growth, starvation, putative phototaxis, and while becoming competent (Rakoczy, 1980).

Consequently, a special metabolic state, or 'mood', of the mould determines whether: (i) growth or spherulation are inhibited in an illuminated well-fed or a hungry plasmodium, or (ii) whether sporulation is triggered in a hungry yet competent one. In the latter case, light acts as a trigger and selects a developmental pathway in a metastable system. Such effects by an exogenous factor are usually termed 'induction' in classical developmental systems. It is obvious though that the inner state of the plasmodium (the 'intrinsic factors') is decisive in whether or not an inductor can be effective. In classical embryology 'competence' has been assigned to a 'sensitive period', and very little is known about why a certain time span is sensitive. In *Physarum* we can define and manipulate conditions which define 'competence' on a time scale, although we are still far from understanding this important phenomenon of developmental biology. One reason for optimism is that the establishment of 'competence' and 'induction' can be uncoupled in *Physarum*. First, the state of competence is developed in the dark; then the plasmodium is illuminated and the developmental consequences can be accurately predicted and analysed once again in the dark. This provides a bio-assay: if sporulation does not occur, whatever changes have been evoked, either by the substrate or by the illumination, they are not

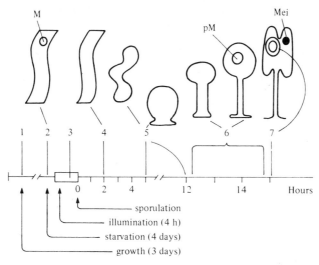

Fig. 51. Some developmental phenomena associated with sporulation (plasmodium–amoeba transition, PAT): 1, initiation; 2, competence; 3, induction; 4, commitment; 5, pattern formation; 6, morphogenesis; 7, cytodifferentiation. M, mitosis; pM, presporangial mitosis; Mei, maturation divisions (meiosis).

sufficient. On the other hand, any positive correlation with sporulation may provide us with a piece of the jig-saw puzzle of the uni-directional differentiation process. Then, for each piece of information it can be asked whether it is essential or not.

Before we discuss some details of competence and induction, let us describe what happens in the developmental sequence (Fig. 51). From the time the light is turned off, it takes about 15 hours for the fruiting bodies to form. Before anything can be seen, the plasmodium has passed a 'point of no return' (PNR). This point has been defined as the last moment when addition of full growth medium can still override (or repress) the new developmental programme and revert the whole creature to a growing mass of protoplasm. After the PNR, about three hours after the end of illumination, the plasmodium is committed to differentiate. The first visible alteration in the precise sequence of events, which takes place in good synchrony (i.e. within ±15 min) throughout the whole plasmodium, occurs at 5–6 h. The plasmodial strands which have been straight until that time, begin to undulate, and these contraction waves cause the strands to assume a zig-zag-like configuration. After a further 5–6 h, these strands break up into beads of 1–2 mm diameter at regular intervals along a vein, into which the protoplasm is collected. However, all beads remain connected by a thin sheet of plasmodial material, covered by a crust of dried slime. The number and size of these nodules depend on the size of the vein and the mass of the whole plasmodium. Although a detailed analysis has not been made, we can infer a patterning process from the viewing of time-lapse films, as the plasmodial network is synchronously transformed into many beads or pre-sporangia (see Fig. 3).

It takes one more hour for these beads to undergo a remarkable example of intracellular morphogenesis (see Fig. 9). The beads become elongated, they rise straight up from the substratum and display rhythmic contraction waves that are obvious in the time-lapse films. These pillars of 1–2 mm height then become constricted, which separates a head from a stalk. During this process pressure increases as the stalk contracts, pressing the endoplasm and its many nuclei into the head structure which is flattened to a disk within 15 min. During this process, the pigment granules become concentrated in the centre of the stalk, which leaves the protoplasm in the head with less pigment and gives it its light-yellow colour. At that time, 15 hours after end of illumination, morphogenesis has been completed. The final event is visible with the naked eye: the fruiting bodies turn black, in a process called melanisation, within about an hour.

As we have frequently referred to the development of *Dictyostelium*, and as its fruiting bodies when finished somewhat resemble those of *Physarum*, it seems worth mentioning that nothing in *Physarum* morphogenesis resembles what has been called 'a fountain running backwards' in *Dictyostelium* (Bonner, 1967). However, there is one formal similarity. In *Dictyostelium*, which has not quite succeeded in becoming a multicellular organism, about half of the cells die as they build up a stalk for the surviving cysts to sit on in the head. Although such 'altruism' is not evident in *Physarum* at the cellular level, a large number of nuclei are degraded during the formation of the fruiting body. At various points during that period (8–14 h after illumination), up to 25% of all nuclei are pyknotic. As it is not known how long the elimination of nuclei takes, an exact proportion cannot be calculated. It is tempting to speculate that nuclei specialised in somatic functions are eliminated, while those capable of propagation survive. All that is known about the survivors is that they become a little bigger than normal nuclei. It should be recalled that nuclear turnover is a characteristic of true slime moulds and occurs in all stages of the plasmodium where about 0.5% of pyknotic nuclei have been observed.

Another obvious cytological event occurs in the yellow heads of the fruiting bodies, at 12–13 h after illumination: a nuclear division, the presporangial mitosis. It is fairly synchronous, although nuclear division seems to move from the periphery to the centre as a wave that takes about 30 min. In this study, Guttes *et al.* (1961) have described metaphase as being different from the metaphase in the growing mould or the starving plasmodium. The nucleoli do not dissolve, and they seem to become distributed unevenly into only one of the two daughter nuclei. In view of our knowledge that the nucleolus contains all the rDNA, and that all amoebae which hatch from spores contain a nucleolus complete with 150 or so minichromosomes, there may be a problem in regulating rDNA gene numbers, if that unequal mitosis is a common event. At any rate, this light-induced mitotic division is not directly coupled to the extensive nuclear elimination, as the latter happens before as well as after that mitosis.

Yet another cytoplasmic event is coordinated with the formation of the yellow head. A large number of vacuoles begin fusing with each other and create a spaceous network of channels in the protoplasm, in which a fibrous material is deposited, which the taxonomists call capillitium, and which aids, by means of explosive hygroscopic expansion, in rupturing the wall of the fruiting body and propagating the spores. This capillitium contains large amounts of calcium which has been stored in vacuoles and in granular form (presumbably as hydroxyapatite) inside the cristae of the mitochondria. Calcium deposits have accumulated to the extent of about 40 granules per mitochondrion during starvation, and half of this disappears one hour after the end of illumination, and none is left in mitochondria. In spores, mitochondria seem generally less active, they resemble those of dormant spherules with respect to the close packing of the tubular cristae and a more electron-dense matrix (Nicholls, 1972).

At about 14 h post-illumination, actually in a similar manner as during macrocyst formation, the protoplasm in the yellow head begins to cleave as vesicles fuse; this leads first to multinucleated units, which are further subdivided until just one nucleus is surrounded by a plasmalemma, i.e. a typical cell has been created for the first time in the plasmodium. Recently, a temperature-sensitive mutant has been described, hts 23 (Burland, Chainey, Dee & Foxon, 1980), in which amoebae cannot divide at the restrictive temperature, nor can spherules or spores be parcelled out by cleavage. This might indicate that these processes are all controlled by the same gene.

Finally, double-layered membrane sheets appear and surround each cell. These walls contain the melanin, otherwise they seem to be composed of branched fibrils of the galactosamine polymer, quite similar to that already encountered in macrocyst formation. As one difference with respect to spherule walls, the walls of the pre-spores, as these cells are called, contain some 2% protein by weight, which consists mainly of three fractions after gel electrophoresis. From this observation one can view these proteins as being a molecular marker specific for the sporulation programme, and we shall come back to that speculation later.

For now we can summarise that sporulation in *Physarum*, as with all major developmental processes in higher organisms, can be dissected into distinct events with well known but not understood terminology, in the following sequence (Fig. 51): attainment of competence, induction leading to a new state of irreversible commitment followed by pattern formation, morphogenesis and, finally, cytodifferentiation. For *Physarum* this leads to the establishment of a dormant, resistant cell – the spore. But there is more to it than that. One consequence is a separation of vegetative functions from those required for the propagation of the species via a unicellular stage. That, of course, is typical for any higher organism, including man, and has been described as the distinction of soma from germ line, i.e. the cells that build up the body as distinct from the germ cells, i.e. the eggs and sperm. In the higher organisms these functions are integrated in the sexual section of the life cycle, which requires meiosis and terminates in the development of unicellular gametes. This occurs in *Physarum* sporangia as well, and under normal conditions the first meiotic division occurs at about 18–24 hours after the pre-spores have formed, and although the time of the second meiotic division is not clearly known, it must have happened within two and a half days after a sporangium has formed, as following this maturation period the spore can be germinated and give rise to a single cell with one haploid nucleus. However, many abnormalities have been observed ranging from pre-spores with several nuclei, all undergoing meiosis but eliminating all nuclei but one, to a single nucleus giving rise to a spore with four haploid nuclei, which, after a mitotic division, can yield eight amoebae from one spore (Laane *et al.*, 1976).

Furthermore, as discussed in the life cycle overview (chapter 1) in *Physarum*, functions of initiating a new life cycle from a single cell (the gamete) can be completely separated from those of meiosis, fertilisation, and recombination of two haploid genomes. A special situation exists in these non-sexual cycles, where plasmodia arise from amoebae by 'selfing' without changing chromosome number or ploidy. In a study of the famous haploid Colonia strain it seems that the nucleus in the pre-spore 'attempts' meiotic divisions after which one daughter nucleus is degraded (Laane *et al.*, 1976).

Finally, a mutant has been described, alc (for amoeba-less cycle) by Truitt & Holt (1979), where a tiny plasmodium emerges from a mature spore instead of an amoeba. This mutant could be the missing link in viewing the life cycle of *Physarum* as a composite of five separate developmental pathways: (1) proliferative nuclear divisions (the mitotic cycle), (2) encystment, (3) morphogenesis, (4) cellularisation, and (5) meiosis, leading to gamete formation and their fusion later on, which are all in operation yet tightly integrated in multicellular organisms, including man, but can be uniquely separated in *Physarum* (cf. Figs. 14 and 51).

Let us now dissect morphogenesis and spore differentiation with respect to metabolic changes and the involvement of gene expression.

Metabolic changes

Although the morphological sequence of events is quite comparable for any mode of obtaining sporulation, a limited period of light after a long period of starvation followed by a distinct point of no return (PNR) and a number of biochemical changes have only been analysed under one set of rigid conditions, which allowed a high frequency of sporulation, albeit not in all strains of *Physarum* tested. Most results have been obtained with the Wisconsin strain ($M3_c$) under the following conditions, carefully specified by Daniel & Rusch (1962) and Daniel (1966). Microplasmodia must be grown in suspension and shaken until the end of the exponential growth phase. Only at the transition to the stationary phase will sporulation occur in macroplasmodia prepared from those microplasmodia (Fig. 52*a*). Recently, it has been claimed that microplasmodia can be harvested in the exponential growth phase and sporulation will ensue if the macroplasmodium is kept on growth medium for one day before transfer to sporulation medium (Chapman & Coote, 1979).

In the classical procedure, a period of 4 days at 21–22 °C in sporulation medium was required, followed by 2–4 h of illumination. Since then, a shorter period of 2 days and extension of the light-sensitive state to 7 days have been also reported (Gorman & Wilkins, 1980). The sporulation medium is a buffered balanced salt solution, which contains as an essential component, either tryptophan or a derivative (niacin, nicotinic acid, NAD or NADP). Usually, niacin is used, which is not necessary for either spherulation or growth. As niacin is a precursor of NAD, many different metabolic effects can be imagined, and it is not clear which of these makes *Physarum* 'competent' to sporulate. Three alternative effects have been suspected: (i) NAD^+ could influence intermediary metabolism by acting as a cofactor of enzymes in the cytosol; (ii) in its reduced form, NADH could regulate the flow of electrons in the mitochondrial electron transport chain and thus influence the ATP level in the cytoplasm; and (iii) as a substrate, NAD could be used to synthesise poly(ADP)–ribose. The respective polymerase is found only in the nucleus, tightly bound to DNA, in *Physarum* as in other eukaryotes (Shall, 1973). This polymer has been shown to react with many nuclear proteins which results in a covalent modification (ADP–ribosylation). This, in turn, could modify nuclear functions – DNA replication as well as RNA transcription. A correlation with the onset of S phase has been observed in the growth state of *Physarum*, but owing to replication–transcription coupling, it is not clear which of the two processes is affected by a high rate of ADP-ribosylation. Recently, a role for ADP-ribosylation has been suspected in the DNA-repair mechanism.

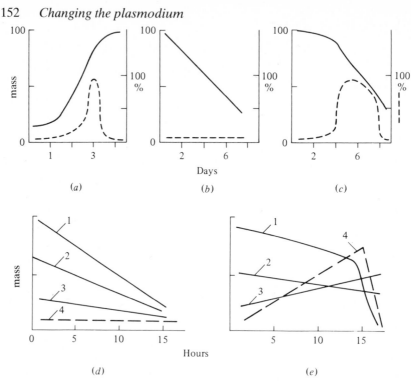

Fig. 52. Correlations between the growth state and incidence of sporulation. (*a*) At end of exponential growth (solid line), transfer to sporulation medium (which defines initiation) leads to best results in sporulation frequency (dashed line). (*b*) During starvation in the absence of niacin, protein content decreases continuously (solid line), no sporulation can be induced by illumination (dashed line). (*c*) In the presence of niacin, initial protein turnover is less pronounced (solid line), and plasmodia are competent during 4–8 days of starvation (dashed curve). (*d,e*) Changes in protein (1), RNA (2), DNA (3) and glucan (4) in a plasmodium incompetent to induction (*d*) and after successful induction of sporulation (*e*).

Although it is not clear how a high level of NAD favours 'competence', some general effects on the degradation of macromolecules during the first two days of starvation and the composition after illumination have been observed (Fig. 52*b–c*; Daniel, 1966). At the end of the 4-day starvation period glycogen has disappeared, protein has decreased by 50%, RNA by 70% and DNA by as much as 80%, irrespective of added niacin. However, while the level of protein is maintained in the presence of niacin in the first two days, DNA drops by one half, resulting in a different ratio of protein : DNA than that in plain salt medium. We have discussed for the growing plasmodium that this formal parameter (protein : DNA about 100) is believed to be essential in maintaining the growth state.

After illumination, the effect of niacin is quite significant: in contrast to the absence of niacin and morphogenesis the level of RNA is maintained. The protein level drops markedly (at the end of the developmental sequence when spores are made), and DNA even doubles in quantity. The last observation indicates that one round of replication is required for sporulation, which actually follows immediately after the pre-sporangial mitosis, as has been confirmed by radioactive thymidine incorporation. Three more effects can only be seen after illumination, in the presence of niacin and in those plasmodia which do sporulate: (i) a transient increase of ATP by 50% at 1 h after light treatment followed by a second increase to twice that level which peaks at 4 h after the end of illumination; (ii) a glycogen-like polysaccharide appears at 4 h, increases until 8 h when the nodules are made, and disappears again by the time the sporangia have differentiated; (iii) exogenous glucose, which at the low concentration of 0.1 mM completely represses sporulation, has no effect from 1 h after illumination onwards. This is a specification of the PNR which was formerly determined by refeeding experiments (p. 147). It has been claimed, not only that the glucose permease system is inactivated at that moment, but that glucose is actually secreted from the committed plasmodium. Until that time, lactate, pyruvate and folic acid are also inhibitory to differentiation. Consequently, the illumination has been suspected to re-orient the carbohydrate and the energy metabolism. Indeed, the other components of the medium, yeast extract (for vitamins) and tryptone (for amino acids), do not repress differentiation, only glucose and some of its metabolites do. This is a clear difference from macrocyst formation, where glucose does not repress the respective differentiation pathway in the dark.

As a final comment on the possible function of niacin, it has been shown in many experiments that the medium or substrate from cultures which have sporulated, while it does not replace the requirement for illumination, can have two effects: it shortens the period in which the plasmodium becomes competent, and it reduces the amount of light necessary to ensure differentiation. In particular, a non-dialysable 'sporulation-control-factor' has been obtained from such conditioned medium which can fully substitute for niacin (Wilkins & Reynolds, 1979). Thus, it may turn out that the plasmodium – which must have a minimum size to become competent – may produce and secrete some factor which, in turn, makes it competent to sporulate. This view is supported by the following observation: if sporulation medium is replaced, after starvation in the presence of niacin, with fresh sporulation medium including niacin, no sporulation occurs after illumination.

While conditioned medium never substitutes for the light in *Physarum polycephalum*, in a related species which also requires light for sporulation, *P. didermoides*, it was observed as early as 1906, in non-axenic culture, that conditioned substrate causes fructification in the dark (von Stosch, 1965).

For *P. polycephalum*, however, it has been shown that sporulation can occur in the dark, at very low incidence, provided it is grown on oat flakes. Furthermore, some species of true slime moulds which happen to be colour-less, sporulate regularly in the dark (von Stosch, 1965, for refs.). At first sight this might indicate that the various pigments that give plasmodia their colour have something to do with the light-sensing mechanism. However, the white mutant of *Physarum polycephalum* (whi) with very little, if any, yellow pigment does require light for sporulation. Actually, it is much more sensitive than the yellow wild type. This might suggest that the massive amounts of pigment serve a protective function in the plasmodium, which would then require a higher light energy than the mutant to react with the true photoreceptor (Rakoczy, 1973).

Although we have no idea what characterises the state of competence, it can be formalised as a quantitative and not a qualitative concept. All ob-servations are consistent with the assumption that a factor, substance A, needs to be transformed into substance B which is required for sporulation and has to reach a threshold level. If it happens without an exogenous trig-ger, no further conditions are needed, and the system shows self-differen-tiation (just as in some tissues in the frog, such as the epidermis). Where an extracellular trigger is required (comparable to the organiser in the amphib-ian gastrula, or illumination in *Physarum*), the system displays 'dependent differentiation', leading to sporulation (or neurulation in the case of amphibia). In this formal model, given by Rakoczy (1973) for *Physarum*, the light energy would promote the accumulation of substance B. If substances A and B are each the result of a complex network of interactions, and the quantity of substance B be affected in many ways, the threshold of substance B, which is required to switch on the differentiation programme and perm-anently maintain the state of commitment, may not be reached in an exact (i.e. predictable) time. This can be suspected from comparing data from different strains or various growth conditions of a particular clone of *Physarum*. Here we encounter for the first time some of the problems which face embryologists, who deal with heterologous cells and their respective *anlagen* inside the eggs. However, looking at the prerequisite for spor-ulation in terms of producing a substance B, we may lose the distinction among competence, induction and determination, but we may gain some understanding from experiments that still need to be done.

Searching for the photoreceptor

What does the light do that is so hard for the experimenter to circumvent in triggering an irreversible differentiation process in *Physarum*? While the previous section dealt with 'competence', this one will be devoted to 'induc-tion'. Since we have already seen that various metabolic changes occur before any morphological changes take place, and as not one mutant is available yet that specifically blocks sporulation, we must rely on general biological observations on *Physarum polycephalum* and the related species *P. nudum* and *Didymium*. Most data have been reviewed (von Stosch, 1965; Daniel, 1966; Rakoczy, 1973, 1980).

Three kinds of experiment in the days of impure culture (reviewed by von Stosch, 1965) demonstrated that at least one effect of the light is long-lived. Firstly, it was found that illuminating the growing cultures reduces the light period that is required, after some days of starvation, to ensure sporulation. Next, it was shown that feeding a starved plasmodium with an illuminated one which had been killed by freezing shortened the illumination period by one-half. Thirdly, if a plasmodium managed to differentiate into a sclerotium even in the light, it could then be germinated and make sporangia in the dark.

In other experiments we have already noted that between two and eight hours of illumination were conducive to sporulation in the Wisconsin strain, and in further studies it was shown that a 30 min illumination period can also suffice. In *P. nudum*, if the intensity of the light that ensures sporulation after 6 h is reduced to 25%, illumination must be extended four-fold, to 24 h. On the other hand, regular shifts of 8 h light and 16 h dark during starvation prolongs the time required for sporulation (as compared to complete darkness before illumination), and short cycles (such as 0.5 min light and 2.5 min dark) have even inhibited sporulation altogether and killed the plasmodium.

In still other experiments, illumination during the first two days of the starvation period did not promote sporulation later on, whereas illumination in the next two days of starvation did so increasingly.

The structural integrity of the plasmodium is not required to preserve the light effect: a plasmodium of *Didymium* was mixed up using a spatula, and sporulation occurred on schedule just the same. Finally, fusion of a competent plasmodium of *Physarum* with an illuminated one results in an inhibition of sporulation altogether, which might indicate that illumination must inactivate an inhibitor of sporulation.

Whatever the light does, it seems to take a long time to be effective. Consequently, any faster metabolic changes (within a minute or less) must be followed by secondary reactions to elicit the slow sequence of events, which include increased migration, inhibition of growth, a change in the colouration of the plasmodium, and finally sporulation proper.

The colour change, together with the fact that some colourless slime moulds do not require light to sporulate, has led to the assumption that the pigments may contain the photoreceptor. None of these pigments has been identified. In *Physarum polycephalum*, they have been variously reported as either three nitrogen-containing polyenes, or five different phenolic compounds, or pteridines, or even carotenes. Pigment research in

Physarum is really in its infancy. Two correlations between pigment and sporulation have been observed: the bleaching mentioned above, and a spectral shift of the pigment of *Physarum* from yellow to brown, which takes several hours *in vivo*, as well as in the illuminated pigment extract. The shift can be effected by light of 400–520 nm, which is very efficient in promoting sporulation.

At first sight the variety of pigments seems puzzling, as a single mutant eliminates them all. Are they all derivatives of one pigment, possibly created during extraction? A clear answer cannot be given because the mutation seems not to be in the structural gene. An aged *whi⁻* plasmodium kept on agar turns quite yellow, but reverts to white when re-grown on fresh substrate.

Whatever the pigments are made of, none has been linked to flavoprotein, and all action spectra seem to indicate that all photosensitive reactions of growth, metabolism, and differentiation are mediated by the blue-light receptor described above (p. 119), which is most likely to be a flavoprotein with absorption maxima at 390, 465, and 485 nm (Schreckenbach *et al.*, 1981) in starved microplasmodia of the white strain. These results are consistent with the classical work by Gray (1938) and a careful study on the pigmented *Physarum nudum* (Rakoczy, 1973). In the latter case it has been shown that for UV light (about 350 nm) an optimum for sporulation requires 200 erg/cm^2, and sporulation is completely inhibited by intensities above about 800 erg/cm^2. Most reliably, sporulation has been induced by blue light in the range of 400–500 nm. At least 2000 erg/cm^2 were needed to yield 100% sporulation. The response curve did not show an optimum, as higher intensities had no inhibitory effect, probably because the yellow pigment screened off the blue light as it turned to brown.

There are two observations which complicate the picture. Sporulation can also be induced by red light at 650–670 nm. Again an optimum response curve has been observed requiring at least 20000 erg/cm^2. At that wavelength no effect has been detected in the developmental pathway leading to macrocyst formation in liquid culture.

In a recent investigation of photoeffects on *Physarum* mobility, it has been suggested that blue light causes negative phototaxis and red light causes positive phototaxis (Hato *et al.*, 1976). As a switch in phototactic behaviour is part of the normal sporulation sequence, we may draw two conclusions: (i) sporulation and encystment are really two different pathways from the beginning of starvation onwards, and (ii) as blue light alone also induces sporulation, positive phototaxis may not be absolutely required for sporulation (Fig. 53).

In addition, green light (540–620 nm) inhibits sporulation, even if it is mixed with stimulatory doses of blue light. It has been speculated that the effects of the blue light can be cancelled by the green light, for which cyto-

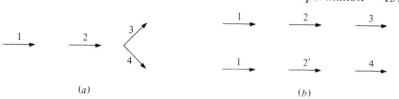

Fig. 53. Two ways of looking at differentiation of the plasmodium. (*a*) After growth (1) and starvation (2), the plasmodium has two options: either encyst (3) in the dark, or sporulate (4) in the light. (*b*) After growth (1) and starvation (2) in the dark, the plasmodium has one option: to encyst (3). In the light, a 'competent' plasmodium (2′) has also one option only: to sporulate (4).

chromes in the mitochondria could be receptors, and somehow disturb mitochondrial function. However, it can be summarised that (i) the right amount of blue light, as well as white fluorescent light in the laboratory and sunshine in nature, will ensure induction of sporulation if the plasmodium is competent, and that (ii) the developmental sequence is remarkably constant and completed within 15 hours with good synchrony.

Therefore, the understanding of 'induction' of sporulation should be brought from the biological to a physiological level if a bioassay were available – make an extract of an illuminated plasmodium, inject it into a dark 'competent' plasmodium and elicit sporulation! Such an assay was described several years ago in a series of elegant experiments. A pigment extract derived from an illuminated plasmodium was dried, dissolved in water, and injected in various aliquots up to a combined volume of 10 μl at many sites into a recipient plasmodium, which was kept in the dark on 2% agar made up in sporulation medium. Sporangia were frequently formed. They looked normal and contained viable spores and gave birth to amoebae (Wormington & Weaver, 1976). The pigment extract was then fractionated and chromatographed into three yellow fractions, and one peak (C) contained the small active principle of about 500 daltons mass. Although one yellow pigment, which showed a spectral shift upon illumination *in vitro*, was also present in the active fraction, its absorption spectrum was different from the action spectrum of the blue-light receptor. In addition, the absence of that yellow pigment fraction from the white mutant indicated that it is not the active principle. Nevertheless, this fraction could be isolated from a plasmodium starved in the dark or even from a growing plasmodium, and after an extensive period of illumination *in vitro* (up to several hours) it became 'activated' and induced sporulation after injection. In the control experiment non-illuminated pigment extract did not. These results indicate that a light-sensitive material is present in the plasmodium at all times, and only a competent one can utilise it in a specific way. Unfortunately, until the substance is identified it is not possible to decide whether a specific differentiation-inducing molecule has been isolated in these assays, as the same authors had previously shown that injection of sodium chloride (NaCl) and ammonium chloride (NH$_4$Cl), also led to sporulation in the dark, although less frequently (Wormington, Cho & Weaver, 1975). And, of course, one must see to inject something, i.e. use some light in the injection procedure, maybe enough to saturate the blue-light receptor.

At this point we can review the many things that happen during the preparation for morphogenesis and cytodifferentiation, i.e. fructification in its restricted sense. Once again, as with macrocyst formation, we can ask whether there is one trigger or many that select a specific developmental pathway which ensures survival and propagation.

The conditions are more stringent than in spherulation, yielding the complex state of competence, which did not show up in macrocyst formation. From the very onset of this pathway, at the transition from growth to stationary phase, there are several options for a plasmodium, provided the conditions are right: to start growing again, to make macrocysts, to die, or to sporulate. Only a narrow time-span is conducive to the latter (see Fig. 52a), and niacin is required in the sporulation medium. Light is an essential factor for the irreversible sporulation process, and it seems to affect many sites, presumably through the blue-light receptor of unknown location. There are changes in the plasma membrane, causing a decrease of glucose uptake and an increase of calcium uptake. A putative light-driven proton transport system (an exchange of K$^+$ for H$^+$ ions leading to a less acid medium, as Daniel, 1976, has deduced from inhibitor experiments) may lead to a transient cAMP increase. On the other hand, the cytosol glucose metabolism is quickly re-oriented (a glucan is made and used up) and less of an unidentified acid is secreted into the medium. This alone could also explain why the medium becomes alkaline, as discussed by Schreckenbach *et al.* (1981). The mitochondria use less oxygen, increase electron transport and make more ATP. The nuclei are triggered to undergo yet another division and replicate their DNA. Although this division occurs late and after morphogenesis is well under way, other nuclear functions are clearly involved, such as RNA transcription.

However, it must be stated that all changes described up to now, whether they are caused by the substrate or the light, resulting in 'competence' and 'induction' respectively, are not sufficient: until the point of no return (PNR) everything can be overridden by one pinch of glucose, even after the light has been switched off. What else is needed to stabilise the sporulation pathway? This raises the long-standing classical issue of 'determination'. While we do not know anything about determination for sure, some microorganisms that have been used to feed plasmodia seem to sense something; *Oscillatoria*, for instance, is attracted to an uncommitted plasmodium yet repelled by one that is committed to sporulate.

We shall now construct a case which indicates that, unlike in macrocyst formation, selective gene expression is an essential step in sporulation. Although most data are on one strain, which has since been shown to produce non-viable spores, and have been obtained from either indirect evidence through inhibitor studies or direct evidence employing what is now considered stone-age technology, the 'story' to be told may stimulate new experiments.

However, as sketched in Fig. 53*b*, one may suspect already that sporulation is not a continuation of a common 'starvation' pathway, where illumination switches off encystment and switches on sporulation. Encystment is a reversible differentiation process (except perhaps for a cell cycle state) that is brought about by many unspecific means. The sporulation pathway, however, requires initiation, a definite (yet unidentified) metabolic state during growth, an extended starvation period and niacin to become competent, as well as red or blue light for induction, and is completely repressed by 0.1 mM glucose until the PNR which will now be discussed in more detail.

A case for gene expression during sporulation

During sporulation overall degradation of protein and RNA exceed by far their respective de-novo syntheses, much more than during macrocyst formation, but it seems that both kinds of molecules must be made after the illumination period to ensure differentiation. Experiments with actinomycin D define a transition period at 2 h after the end of illumination (Sauer *et al.*, 1969*a*), which is quite similar to the PNR defined by refeeding or glucose repression (if 1 h is allowed for actinomycin D to become effective). Although a new permeability barrier to the drug could not be shown, nor rigorously excluded, and transcription may still be required at a later time, the main argument here is that this inefficient inhibitor of RNA synthesis does prevent sporulation even after the induction by the light. Cycloheximide, the very efficient and specific inhibitor of protein synthesis in *Physarum*, blocks development of sporangia if added until 7 h postillumination, at a time when the morphogenetic process is getting under way. Thereafter, this drug does not effect the completion of morphogenesis, although the final marker, melanisation at 15 h, can still be prevented as late as 13 h after illumination. In this instance, the drug probably prevents de-novo synthesis of polyphenoloxidase and phenolase, two enzymes required at the very end of the developmental sequence to catalyse the formation of the black pigment that is incorporated into the fruiting-body structure and into the spores (Chet & Hüttermann, 1977).

In a more direct approach, radioactivity incorporation studies into RNA revealed a somewhat higher incorporation of uridine, when the label was continuously present during the 4 h illumination period and 4 h thereafter. A five-fold increase in specific activity in the RNA of illuminated cultures has been detected, where incorporation of inorganic ^{32}P into RNA has been measured. Although light might have affected permeability differentially, it seems justified to say that RNA transcription does not decrease in the plasmodium as it gets ready to sporulate (Sauer, Babcock & Rusch, 1969b). Pulse-chase experiments and sucrose-gradient analysis clearly show that most label of the heterogeneous nuclear RNA pattern disappears over a 4 h chase during the starvation period. At that time rRNA is the main labelled species in the cytoplasm. The picture looks different when labelling is done during the 4 h illumination and the first 2 h thereafter. While the pulse-labelled pattern in the nuclei looks quite similar to the dark control, there is always less label in rRNA and more label in putative mRNA. It could be, of course, that the larger proportion of this non-ribosomal RNA in the illuminated plasmodium is a degradation product of rRNA; however, several points argue against that. Firstly, among the mRNA labelled during the light, there is relatively more heavy RNA than in the dark control, which can appear as an extra peak of 10–20 S in the RNA profile obtained during the first 2 h after the illumination. Secondly, DNA–RNA hybridisation of total labelled RNA with DNA (under conditions which detect only transcripts of repetitive DNA sequences) have indicated that at 2 h after illumination the RNA can saturate more DNA which cannot be displaced by unlabelled RNA from a non-sporulating starved plasmodium (Sauer & Rusch, 1970). Thirdly, a preliminary estimate of radioactive poly(A)$^+$RNA indicated a three-fold increase in the illuminated plasmodium after *in vivo* labelling, and endogenous RNA polymerase B activity in isolated nuclei, almost completely absent after 4 days of starvation, is reactivated during 4 h of illumination plus 2 h in the dark (R. Simon, unpublished). As a final comment, in plasmodia that are competent to sporulate, a modified RNA polymerase has been extracted in a soluble form. This form of the enzyme elutes at a higher salt concentration from the column, and its activity is more sensitive to salt than RNA polymerase B, but it is sensitive to α-amanitin (Hildebrandt & Sauer, 1976a).

Even if these results can be confirmed with improved methods, it is clear that transcription of non-ribosomal RNA alone does not commit the plasmodium to sporulate, since these new RNA fractions do not prevent the mould from reverting to growth in the appropriate medium. If the new RNA has to be utilised soon after transcription, say at the PNR, an irreversible commitment could occur as a consequence of renewed gene expression. This may be the case, as a heat shock of 30 min at 37 °C (which has been shown to disaggregate polysomes, thus inhibiting the translation of mRNA) (Schiebel *et al.*, 1969) also has an effect on sporulation. During 5–9 h after illumination, abnormal morphogenesis has been observed to interfere with the formation of stalk and head. An inhibition of sporulation altogether occurred in about 50% of treated plasmodia when the heat shock was applied during illumination and the following 4 h. No effect was seen until 10 h before illumination and 10 h post-illumination. While these observations are consistent with necessary protein synthesis, the requirement of transcription for immediate translation can be seen from another series of experiments involving actinomycin D. This drug does not interfere with amino-acid incorporation for at least 6 h during growth, but, as we have seen above, after 24 h of starvation a drastic inhibition takes place in microplasmodia. After 4 days of starvation, a macroplasmodium has again become more independent of transcription, and only 15–20% inhibition of amino-acid incorporation occurs over 5 h. However, when actinomycin D is applied 1 h after illumination, inhibition of protein synthesis reaches 50%. Later on, after commitment (when the drug has been added at 5 h post-illumination) only 20% of amino-acid incorporation is inhibited (Sauer, Goodman, Babcock & Rusch, 1969).

All observations so far indicate that after commitment some freshly made RNA is translated into protein. Indeed, the polysome pattern shows characteristic changes: it is degraded during the light treatment but is restored after the point of commitment and contains even bigger polysomes than in the starved plasmodium kept in the dark (Jockusch *et al.*, 1970). It seems as if illumination is a shock for *Physarum* and amino-acid incorporation – in contrast to RNA transcription – clearly decreases during the light period. If proteins are labelled for about 5 h after the light treatment (i.e. before morphogenesis has begun) the labelled protein that is extracted in an aqueous buffer contains relatively more large polypeptides. Another protein fraction, rich in structural protein, has been solubilised in acetic acid. It shows two distinct differences from the committed plasmodium. (1) A protein fraction of low mass (15 000–25 000 daltons) is less labelled. (It resembles histones with respect to its mobility in a gel.) (2) Three protein bands in the range of 50 000–80 000 daltons are selectively labelled in the sporulating mould (Jockusch *et al.*, 1970). It may be a coincidence, but two of the three protein bands detected in the spore wall by McCormick *et al.* (1970) comigrate with two of the labelled bands of the structural protein extractable only from a plasmodium committed to sporulate.

Although a confirmation of these results is badly needed in other strains of *Physarum*, it could be that, due to the extensive starvation period, in combination with the 'shock' to protein synthesis exerted by the light, the selective expression of those genes specifically involved in an irreversible differentiation process, can be analysed against a very low background of household functions. These observations on putative gene expression regulation during sporulation are summarised in Fig. 54.

Little is known about the cytodifferentiation of the spores aside from the cytological changes during their formation in the yellow head of the sporangium (Aldrich & Blackwell, 1976) and other fine structures that have

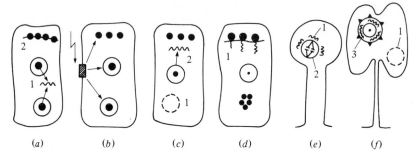

Fig. 54. Illustrating altered genome expression during sporulation. (*a*) Nuclei of a starved plasmodium make mainly rRNA (1), and some polysomes (2) are in the cytoplasm. (*b*) Illumination leads to degradation of polysomes and affects nuclei (arrows). (*c*) Some nuclei become eliminated (1), others make mRNA (2). (*d*) Polysomes appear as the fresh mRNA is translated (1). (*e*) During morphogenesis, fresh proteins (1) occupy the cytoplasm, and presporangial mitosis (2) takes place. (*f*) More nuclei are degraded (1) as spores with some fresh protein as part of their walls (3) differentiate within the sporangium.

been determined with the electron microscope and which are of great value for taxonomists (Hohl, 1976).

Only after a synchronous mitosis takes place do spores form. This brings us to a brief discussion of the mitotic cycle events during sporulation.

The state of the mitotic cycle

We have already mentioned that during the period in which a plasmodium becomes committed to sporulate, the mitotic cycle lengthens to 24–36 h (Guttes *et al.*, 1961). It seems that a G_1 phase does not occur during starvation, as the DNA content remains at about 1 pg per nucleus, which is characteristic for the G_2 phase. The nucleolar reconstruction takes several hours, however, clearly more than the 45 min during the growth phase, and the duration of S phase and other pertinent questions, like the temporal sequence of replication, have not been studied. Even after the light-induced synchronous nuclear division in the young fruiting bodies, DNA synthesis follows nuclear division immediately (Fig. 55). It is clear that these mitoses occur in the absence of growth, at a time of extensive protein degradation, and none of the arguments discussed concerning the triggering of mitosis, which were summed up in the hour glass model, can apply here, where nuclear divisions take place in a plasmodium which is off balance (unless it is in reality suspended in a network of carefully balanced anabolic and catabolic reactions). In this respect, there may be similarities to multicellular organisms, where a true exponential growth phase never exists, unlike the balanced *Physarum* plasmodium.

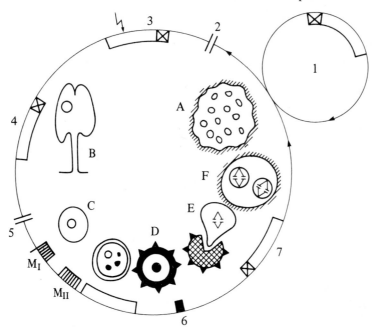

Fig. 55. Variations in the mitotic cycle during the life cycle of *Physarum*. Fast proliferative cycles (1), where mitosis (×) is followed by S phase (white bar) in the
plasmodium (A); slow cycles during starvation (2); putative quantal cycle and
commitment to sporulation by illumination (3); light induces presporangial mitosis
(4) in the sporangium (B), followed by a maturation period (5) of the spore (C); two
meiotic divisions (M I, M II) and G_2 phase arrest (6) lead to the mature spore (D)
with one haploid nucleus; upon germination of amoebae (with open mitosis) (E)
proliferative cell cycles set in (7), following amoeba–plasmodium transition; fast
proliferative nuclear cycles occur in the young plasmodium with closed mitosis (F).

A puzzling phenomenon is the preferential fructification in the middle of
the night, optimally at 2 a.m. (von Stosch, 1965). We can rationalise this
finding by assuming that the end of illumination may be less important in the
developmental programme than the beginning. The end of the day is too
close to trigger the nocturnal fruiting, but the beginning of the day, say at 7
a.m., allows just the time required under Daniel's rigid programme which
adds up to 19 h or 2 a.m. (4 h light plus 15 h until melanisation).

Going back to this rigid regimen, a mitosis has been observed at 92 h of
starvation, just 2 h before illumination. Therefore the nuclei in the first 2 h,
the obligatory light exposure time, have still been in S phase. Experiments
interfering with this DNA replication with either bromodeoxyuridine or
hydroxyurea inhibited the whole developmental programme (Sauer,
Babcock & Rusch, 1969c; Sauer, 1973). Maybe this round of replication (and
transcription coupling?) is an essential step in becoming 'competent' to the

illumination (Fig. 55). Some DNA synthesis has also been observed at 7–8 h after the illumination, when the beads, the *spore-anlagen*, transform into the pillar stage, which is just before heavy nuclear degradation sets in, and before mitosis. The light forces that mitotic division at 13 h after illumination (or 17 h from the onset of illumination), which is certainly earlier than the 24–36 h interphase of a hungry plasmodium in the dark. Many nuclei may not get ready in time to divide and thus become pyknotic and are eventually eliminated.

While these observations of light sensitivity and DNA synthesis are vague, generally at least 50–70 h must elapse before the light can trigger sporulation. This interval is just long enough for two cell cycles after the transfer and synchronisation of microplasmodia at the end of the exponential growth phase into non-nutrient sporulation medium. The imbalance of competent macroplasmodia can also be deduced from some changes in nuclear protein: the histone : DNA ratio (1.1 in growing nuclei) has dropped to 0.8, and among the non-histone proteins four additional bands have accumulated in a competent plasmodium, and they disappear within 24 h if it is re-balanced by the addition of growth medium (Walker *et al.*, 1980). At the time of the pre-sporangial mitosis, even more drastic changes have occurred with respect to heavy loss of DNA and protein. Clearly, the rules of the 'Kern-Plasma-Relation', which postulates a mass increase before mitosis, cannot apply. What else triggers this nuclear division?

We simply do not know, but an almost universal observation in animal cells is that they have to divide before any kind of stable or 'terminal' differentiation can occur, and the whole concept of 'quantal' cell cycle has evolved around that phenomenon (Holtzer *et al.*, 1975).

Mohberg (1977) has done an interesting measurement on the DNA content of spores after their formation and maturation. The DNA content of the mature spore is 2C. This would imply that spores are arrested and can remain for years in G_2 phase. This is quite different from the putative G_1-phase arrest during macrocyst formation (if these DNA measurements can be confirmed in other strains) and almost all other examples of cytodifferentiation, where the cell cycle arrests clearly after mitosis yet before S phase.

Some curious exceptions do exist which have been summarised by Prescott (1976). For reasons unknown, nuclei in the epithelium of the rabbit ear can arrest in G_2 phase. A remarkable similarity to *Physarum* spores, however, exists in insect embryos, where the switch from the ultra-fast nuclear cycles (<10 min in *Drosophila*) during the cleavage stage to slower cell cycles (about 1–2 h) at the blastoderm formation (another case of cellularisation in a plasmodium) is brought about by the 'invention' of a G_2 phase. Another telling example is some plant meristems, which can also arrest in G_2 phase.

If a quantal cell cycle, or the round of DNA replication connected to it, is an integrated part of a differentiation sequence, *Dictyostelium*, which does not include a mitosis in its developmental programme is either an exception, or it is not a case of stable 'terminal' differentiation, even if a lot of the aged amoebae must die in the process. This gives us a clue to look at a really 'terminal' event in the life of a plasmodium of *Physarum*: death, as a consequence of aging.

Some strains have become extinct and nothing tried has been able to stop the aging phenomenon. One particular condition pertains to prolonged culturing on agar. Despite frequent subculturing, some strains do not survive more than a few months, while in shaken suspension they do so for years and maybe indefinitely. In this case, an enormous enlargement of the nuclei has taken place. It is unknown whether they omit to divide after DNA replication, or whether they fuse with each other, but they contain a multiple level of DNA (McCullough *et al.*, 1973).

A similar observation has been made on the close relative *Didymium*, where a very elegant experiment has been done by Clark & Hakim (1980*a*, *b*), that can be adapted to many other questions: an aged plasmodium with large nuclei has been fused with a vital one, containing normal, i.e. small, nuclei. A porous filter is put between the young and the aged plasmodium, small enough to withhold, and thus eliminate, the large nuclei from the fused plasmodium. The result is striking: the fused plasmodium containing young nuclei plus mixed old and young cytoplasm, can be subcultured again and again; it seems to have become rejuvenated.

With respect to sporulation, one final point must be raised, taking into account that, unfortunately, it has been so hard to get plasmodia to make fruiting bodies reliably on an accurately timed schedule. In many experiments this process has taken on an air of randomness, allowing only the assignment of some vague 'probability' for differentiation really to occur. It is worth mentioning that random transitions are also discussed in stochastic models of cell division, as well as in cell differentiation controls, and in sporulation of some bacteria as well (Sauer, 1980, for review).

An even more serious problem for *Physarum* is that after prolonged culturing in the vegetative growth phase its tendency to sporulate tends towards zero. Many strains, after some years of propagation in the laboratory as shaken microplasmodia, do not sporulate any more. One reasonable explanation has been that a selection process has taken place and that those genes for differentiation, being not required for growth, might have become selected against. As with all conclusions from negative results, they must be corrected as a positive result is presented. This has just been done in *Physarum*, and a strain that lost its ability to sporulate over a period of five years, regained it in just one week after growth on oat flakes, still the best health food for *Physarum* (Hosoda, 1980).

While this result clearly argues for a stable genome of some essential developmental genes in plasmodia, preliminary characterisation of DNA extracted from the yellow heads of young sporangia reveals quite different elution patterns from AT-specific adsorbents as compared with DNA from a growing plasmodium (Wick, 1981).

Since the oat-meal experiment stresses the importance of the substrate in sporulation, and the effect of light indicates other conditions remote from the genome, while the work with RNA synthesis implies genome regulation during sporulation, let us now attempt to integrate all three aspects before we turn to the microscopic forms of *Physarum*.

Some thoughts on the regulation of irreversible differentiation

In a very thoughtful appendix to the classical analysis of sporulation in *Physarum*, Daniel (1966) has discussed this differentiation process as the consequence of two interrelated patterns of metabolic change. They manifest themselves as changes of the flux of energy and the flux of matter through the plasmodium.

Most of these changes occur in the closed system of the starving plasmodium, as it becomes changed from one stable steady state (exponential growth) to another stable state, utilising only endogenous energy sources, leading to spherulation and to yet another stable state, the spore, which requires exogenous radiant energy (particularly blue light). The latter pathway can be initiated only after most endogenous energy sources have been used up and the cell has become 'competent' or receptive to an exogenous stimulus. In this respect the light is both a trigger along the sporulation pathway and the 'information' that is passed on the nucleus, which is considered a memory store of previous environmental, mainly adverse, conditions. The dissipation of the light energy then leads to a specific result, the spore, which is fit for survival. Sporulation requires not only a switch in the metabolic pattern but also the expression of different genes. Clearly, the exogenous signal elicits metabolic changes before the genome becomes involved, which is dictated by the low speed of gene expression in eukaryotic cells and the long life of their mRNA molecules.

In preparation for the irreversible transition of the plasmodium to the sporangium, without nutrients being added, the organism must minimise the conversion of its organic substance into energy in order to maintain the living state. It could do that by (i) specialisation of pathways already in operation during the growth state, by selecting one which is more efficient along the route of differentiation, or by (ii) extensive degradation and re-cycling of small carbon skeletons into new matter.

Higher organisms have evolved sophisticated regulatory mechanisms that modulate and optimize enzyme activity, and glycogen phosphorylase, which is activated by covalent modification (phosphorylation) via hormone signal, may serve as an example. This mechanism does not function in *Physarum*, as we have already seen. Nevertheless, a 'signal' is also apparent in *Physarum* – the light. For it to work, only very little, if any, catalytic activity is needed. Activation of the photoreceptor may be, in principle, similar to a hormone binding to its receptor or the attachment of a virus to the cell membrane before its penetration into a cell. As a general mechanism, then, the transduction of a signal by some second messenger is essential in generating some sort of 'information' which alters both existing metabolic patterns and gene expression. And that goes hand in hand with a change in the 'flux of matter'.

All organisms maintain an ordered heterogeneity of their constituents by compartmentalisation, utilising biomembranes. The biomembrane is asymmetric in structure, having different catalytic protein molecules on either side. This helps in maintaining an ordered steady state, which is achieved either by an active barrier or by setting up gradients of every size of molecule, including water.

Where the enzymes are not limiting, fluctuations in the substrate will regulate the flux of matter. When an enzyme becomes limiting, either by modification (a change in conformation or the availability of cofactors) or by covalent alteration (limited proteolysis, phosphorylation) or degradation, a new steady state can be established. It can manifest itself as a new phenotype, say a sporangium. This can also result from appearance of new enzyme (by activation or by de-novo synthesis). These protein changes, when they assume a stable steady state, either epigenetically (altering existing gene products) or genetically (increase or decrease of genetic readout and processing of that information, eventually in form of an altered protein pattern) require biosyntheses of some sort, i.e. substrates plus energy.

In differentiating *Physarum* plasmodia, membrane fluxes change, including massive membrane production in spherulation as well as in sporulation. This provides for the distribution of both prerequisites which are derived from degradation processes and light absorption. Degradation of large molecules and recycling of small molecules can be more efficiently done, and the transport of an amino acid (as an energy source for gluconeogenesis or as a building block for new proteins) can become site-directed, not by secreting hydrolytic enzymes to the outside of the cell, but by forming more membrane-covered endogenous compartments (autophagosomes). If membranes are the site for convertion of light energy into ATP, either as a light-driven proton pump of the plasmalemma, or 'pulling' the electron transport chain in the mitochondria (a claim that still needs to be confirmed), then the increase in mitochondrial cristae and the many cytoplasmic membranes, vesicles and vacuoles, eventually leading to the cellularisation of cysts and spores, makes 'sense'.

One other high cost of energy, along with the flux of matter and membrane work, which becomes so obvious by looking at a plasmodium as it makes sporangia, concerns the distribution of water. Overall, there is a dramatic shift from the plasmodium, with 80% of its weight being water, to the spore with 80% of its weight not being water. Time-lapse films reveal many contractile vacuoles, madly pumping out that vital solvent, until in the electron microscope the cytoplasm and the mitochondrial matrix look 'inactive'. In addition, during the morphogenetic process, motility, as seen in the shuttle streaming and the erection of the pillar of the sporangia and its constriction to segregate a living head from the dead stalk, ceases as more and more water is lost from the mould. One can argue that dormancy is solely caused by the loss of water, and indeed, distilled water is an excellent rejuvenating medium for a spore: it comes to life, grows and divides without any added food source.

At any rate, we can conclude that differentiation in *Physarum* is associated with a reorientation of the metabolic state which, in turn, is associated with a severe stress situation. This reminds us of three general points, one of which has been given serious consideration in *Dictyostelium* development; the others refer to some puzzling observations in all classical experimental embryology.

In the first case the question has been raised: what is the essential component of a new developmental programme? Is it the restructuring of existing metabolic patterns or the selective gene activation at the level of genome transcription? These alternatives can be described as the 'epigenetic' view of biochemists like Wright (1973) or the 'genetic' view of Sussman (1966, 1976), to name two people interested in *Dictyostelium* development. While the latter view seems to be more fashionable at the moment, the alternatives are unresolved, in *Physarum* as well.

The second point is best illustrated in a classical statement by Seidel, an eminent developmental biologist. 'The closer you bring a young embryo, or a blastomere of it, to the brink of death, the more developmental capacities it will show.' Indeed, it is often seen that at decisive stations in the developmental pathways, cell death does occur. This is seen in the morphogenetic carving out of a hand or a wing from a plump bud, in the general overproduction of motor neurones, over 90% of which get eliminated where contact to a limb is not established, and very early insect development, while the germ *anlage* experiences 'determination' (Sauer, 1980 for discussion). Therefore, nuclear elimination during sporulation of *Physarum* may be quite normal, yet just as mysterious as in these other examples.

The third point concerns 'developmental-information transfer'. This has eventually led to a rethinking of the classical 'organiser' concept, originally proposed by Spemann (Spemann & Mangold, 1924). In its original form the 'inducer' has been thought to provide developmental information that is not present in the reacting embryonic tissue. In the concrete example, the archenteron roof induces the overlying ectoblast to become neural ectoderm.

In view of the many unspecific, even inorganic 'inducers' that could mimic the neuralisation of isolated ectoblast and the 'autoneuralisation' requiring no external inducer in some amphibian embryos, as well as under many other experimental conditions, the inducer is no longer considered as providing information, but rather as a 'trigger' that evokes a developmental pathway that is open to the reacting, competent tissue. In other terminology, the inductive reactions in embryonic systems are now generally viewed as 'permissive', and not as an 'instructive' event. This has a sobering consequence for the enormous stability of the determined state which would have to be explained by the 'instructive' theories as a Lamarckian event, as that state is 'propagated' or 'inherited' over many cell generations (Sauer, 1980 for discussion).

For sporulation in *Physarum*, illumination acts in triggering rather than in information transfer, just as for most embryonic systems and for the effect of many hormones and growth factors in mammalian tissues. Light normally provides the signal; however, it can also arise by 'self-triggering', if the substrate contains oat flakes or if sodium chloride is injected into the mould. This conclusion stresses once again the importance of 'competence' the metastable state required in this irreversible differentiation pathway.

Although, as already referred to above, the natural trigger is more reliable, a sense of transition probability is left in one's mind, when one developmental state is abandoned and another one is gained that is both stable and irreversible. Both mechanisms, the epigenetic metabolic patterning and genome expression, seem to be involved in the first irreversible differentiation process we have discussed for *Physarum* plasmodia. However, there may be yet another mechanism which results in an alteration of the genome. With respect to the structural lability of the genome (as seen from movable elements, the transposons in bacteria, endogenous viruses, rearrangements in the immune genes of lymphocytes and the inability of a nucleus from a fully differentiated tissue to support complete development in an enucleated egg), it could well be that qualitative or quantitative changes of the DNA (transposition, over- or under-replication) or modification (like methylation) play a role in 'commitment'. Although this is purely speculative, we know where to look for a goodly portion of synchronous developing matter in *Physarum*, and the methods have become available for a thorough scrutiny of its DNA.

One last argument concerns the similarities and differences between macrocyst formation (which is readily reversible) and sporulation (which is irreversible). Is the latter just the continuation of the former, or do we encounter two alternative, mutually exclusive, developmental programmes? This is not a trivial matter, although the available data may be too trivial to reach a decisive answer. There are two ways of looking at this question.

(1) An impressive body of similarities can be cited for the two resting life-forms of *Physarum*, the macrocyst and the spore, although they differ markedly in size and shape. Similar characteristics and contents include: reduced oxygen consumption, extensive turnover by degradation, de-novo synthesis of some catabolic enzymes like proteases, a phosphodiesterase and glutamate dehydrogenase, production and secretion of slime and wall structures, dehydration, accumulation of calcium deposits and morphological changes in the mitochondrion, a changing pattern of non-histone protein, and a reduction of all DNA-binding phosphoproteins, an overall reduction of protein biosynthesis, and RNA transcription. Consequently, a spherule has been looked on as a 'spore', lacking phenol oxidase to darken the wall and some structural protein in the wall, and lacking the mechanism for meiosis (Gorman & Wilkins, 1980 for discussion).

(2) The remarkable similarities are all due to starvation and the remaining differences may allow us to discriminate essential developmental requirements for two different pathways of the vegetative plasmodium. Then microcyst formation becomes one of several other pathways open for the microscopic life-form of *Physarum*.

Here are some of the differences specifying the putative sporulation versus spherulation programmes: requirement for niacin; attainment of competence; the repression of the whole differentiation sequence by glucose; the induction by light, involving the blue-light receptor (but also in red light, in contrast to the light inhibition of the spherulation programme and inhibition of sporulation by green light); the putative quantal cell cycle in attaining competence and another one following induction by light; nuclear elimination yet doubling of DNA, and net synthesis of polyglucan during sporulation; the state of irreversible commitment; morphogenesis (in addition to cytodifferentiation also seen in cysts); the involvement of selective, possibly new gene expression; and finally, the putative cell cycle state of G_2-phase arrest in spores and G_1-phase arrest in cysts.

As view (2) demonstrates, an impressive number of events occur only during sporulation (see Fig. 56). These are quite some differences. However, here we face the problem of which processes are due to the alternative developmental programme and which are due to illumination but insignificant for development. Clearly, what is needed to solve this problem is to have sporulation-specific mutants, now that we have so much descriptive data of all sorts. There is one heat-sensitive mutant (hts 23) which interferes with amoeboid growth, spherulation and sporulation alike, probably by blocking 'cellularisation' in the plasmodium as well as cell division in the amoeba. Consequently, this gene is pleiotypic and an example of a general function that is not specific for only one developmental programme, but for all. Therefore, the extensive membrane formation may be added to the list of unspecific developmental functions.

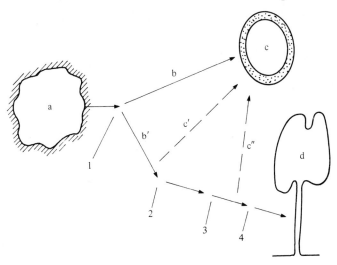

Fig. 56. Sporulation, an irreversible and essential developmental pathway. A growing plasmodium (a) succeeds in sporulation (d) only if it experiences initiation (1), competence (2), induction (3), and commitment (4); whereas starvation alone (b) or all events up to commitment can still lead to spherule formation (c, c′, c″).

While we have to wait for genetic evidence to complement our knowledge of the transition of the plasmodium into a sporangium, the next chapter outlines a situation which is just the opposite, where a complex body of genetic evidence is emerging in a complete vacuum of biochemical data. With this statement we are ready, if not competent, for a description and the genetic analysis of the microscopic lives of *Physarum*.

7

The microscopic lives of *Physarum*

The spores in a fruiting body are many. They are dormant and very resistant. Spores can hatch and release one, or sometimes two, motile cells that are vital and versatile creatures (see review by Gorman & Wilkins, 1980).

The first sign of germination following rehydration is cytoplasmic movement, as revealed in time-lapse films. Then the spore wall is partially lysed, or it just tears open, and the cell squeezes itself through a narrow slit in the spore wall which is left behind. Not all spores germinate, and viable progeny can vary from a fraction of 1% to 90% of the spores plated out on agar in water or buffer. This could be for two reasons: either the nuclear elimination process initiated in the young fruiting body has continued to various degrees in the maturation process, or we have not learned enough about environmental factors that lead either to hatching or to maintaining the dormant state.

The emerging cells have several options; they are truly multipotent. Although unable to turn right back into spores, they can: (i) proliferate as amoebae, probably indefinitely, (ii) transform into flagellated swarm cells and swim around, and (iii) under a variety of adverse conditions, turn into cysts (microcysts, to distinguish them from the multinuclear macrocysts or spherules). Moreover, they can become 'gametes' which have three different pathways to select from, two of them leading to the macroscopic plasmodium stage. Firstly, two haploid gametes can fuse (cytogamy), and after their nuclei have also fused (karyogamy) a plasmodium forms by repeated nuclear division. Secondly, two gametes can fuse, yet when their nuclei fuse into a common mitotic spindle, cell division takes place and two diploid amoebae are created, which can multiply but not change into a plasmodium. Thirdly, the nuclei of one gamete can begin to proliferate in the absence of cell division; thus, the gamete turns directly into a plasmodium, which is just as haploid as the amoebae they produce after a full life cycle. The latter case, the selfer, is an excellent example to demonstrate that two very different phenotypes can be constructed utilising the same genotype. The selfer becomes, then, a fine system in which to look for 'gene-batteries' that are controlled in a mutually exclusive manner, specifying alternative developmental pathways. The beauty of this system for genetic analysis lies in the fact that in the third situation both states, amoeba and plasmodium, are

172

haploid, so that recessive mutations can be immediately recognised, without the need to outcross the wild type allele in a diploid state. Some conditional mutants are available that reversibly block the transition of amoebae to plasmodia, thus synchronising the whole population at the transition point, and culturing techniques are now good enough to provide sufficient amounts of material for chemical analysis.

However, genetic analysis has prevailed in recent years and many methods of microbial genetics have been adopted to study large populations of amoebae (10^7–10^{10} individuals), and very rare mutants, occurring at a frequency of only 1 in 10^9, can be detected. Clever tricks for screening and selecting mutants have been employed, as reviewed by Haugli, Cooke & Sudbery (1980), which, together with chemical mutagenesis, allow one to 'fish out' any mutant, if one knows what to look for. That seems to be the main problem with the most important and unique situation in *Physarum*: the transition of the microscopic to the macroscopic life-form which, at least formally, may be following the same principle in moulds and man.

We shall settle first for some minor problems, where a combination of microbial genetics and classical Mendelian genetics has given some clear-cut data. However, in the course of the cell transformation we shall observe some glimpses of a remarkable flexibility in the genome organisation at the mating type locus. Before we do that, let us analyse the various reversible differentiations of the amoeba–flagellate–microcyst state.

Amoebae

No doubt, amoebae differ from plasmodia in many ways. However, there are similarities as well. Amoebae are small (15–20 μm in diameter). Their nuclei and mitochondria look like those of plasmodia. The chromatin is organised as nucleosomes, each containing just as much DNA as those of a plasmodium. The organisation of the DNA with respect to numbers and site of tRNA genes and rDNA in minichromosomes is also the same, as is the histone composition. However, the non-histone protein fraction (NHP) is completely different. It has been suspected by Goodman (1980) that this reflects the different growth conditions of amoebae and plasmodia at the time those data were obtained. The cytoplasm contains an elaborate endoplasmic reticulum, Golgi apparatus, one or two contractile vacuoles, digestive vacuoles, various other vesicles, and typical centrioles (just as in protozoa and animal cells). The latter are completely absent in the plasmodium. There are neither pigment granules (putative spherosomes) nor yellow pigments in amoebae. No slime is secreted by amoebae, and it has been very difficult to demonstrate anything on the outer surface of their plasmalemmae. However, a thin coat of glycoprotein and sulphated polysaccharides (slime?) has been indicated by staining with Alcian Blue and

Ruthenium Red, respectively (Dykstra & Aldrich, 1978). The movement is amoeboid and no shuttling is observed, yet the motile apparatus (actomyosin) is present and the motive force is created by contraction at the tail end, pushing endoplasm into lobose pseudopodia at the front end of the amoeba. Several mutants are known that manifest themselves in altered colony formation, but none is immobile; actually all seem to move faster than the wild type (Jacobson, Johnke & Adelman, 1976).

Amoebae grow best with microorganisms that they feed on by phagocytosis. They like to graze over a lawn of living or killed bacteria as they grow and multiply by typical cell division. They can be grown from one individual (i.e. as a clone) while they form a plaque in the lawn, which can be seen with the naked eye. They have been successfully grown in liquid medium, ingesting nutrient by pinocytosis. Under these axenic conditions growth is about four times slower compared with the four to eight hours generation time in two-membered cultures (McCullough & Dee, 1976). Some strains have become adapted and grow faster in the liquid medium. From those, two genes (*axe* 1 and *axe* 2) have been defined which are now being crossed into other strains of interest (McCullough, Dee & Foxon, 1978). However, amoebae can now be grown in a medium that supports both amoebae and plasmodia alike (Burland & Dee, 1979). This has led to an important clarification as to the degree of overlap of gene-batteries required to sustain an established amoeba or plasmodium. At first, when conditional mutants (ts mutants that grow at 22 °C but not at 28 °C) were analysed in different media, most *ts* genes seemed to be associated with only one of the two states. Testing them again in the common medium, most *ts* genes were proved to be expressed in amoebae and plasmodia alike. Also, some genes for known products are active in both states, e.g. those for cycloheximide resistance (specifying a modified ribosomal protein) and thymidine kinase. Both of these genes can be efficiently mutated (at a rate of about 10^{-3}), and out of 25 TK$^-$ mutants 24 are in the same gene and one is not (Lunn, Cooke & Haugli, 1977). Some enzyme patterns are similar, such as a group of proteases; others differ, such as the RNases. In the former example, some proteases are excreted by the amoebae, when well-fed in liquid medium. This does not occur in plasmodia nor in slower growing amoebae, and may, of course, affect their environment, including their prospective mating partners (Haars *et al.*, 1978). In the other example, whereas plasmodia have several RNases, only one is found in amoebae.

One comment can be made on the cell-cycle state of amoebae. After hatching, within 1–1.5 h, a fairly synchronous nuclear division takes place. As we remember, mature spores seem to be arrested in G_2 phase, allowing for their division after hatching within a time span that is even too short for a full round of DNA replication. One further similarity between the two states is the absence of a measurable G_1 phase, even though the generation time of the amoebae is very long.

One drastic and probably very essential difference between the two states is in mitosis itself: it is typically 'open' in amoebae, with a bipolar spindle, consisting of two half-spindles, each complete with a proper centriole at its pole and connecting to one set of the chromatids, after the nuclear membrane has dissolved. On the other hand, plasmodia have a closed mitosis with a persisting membrane around the nucleus and an intranuclear spindle (Goodman, 1980). Nucleolar reconstruction looks about the same in both cell types, which is consistent with both having extranuclear ribosomal genes and minichromosomes. One mutant (hts 23), already mentioned as interfering with differentiation of the plasmodium, tells us that to make a plasmodium it is not enough to dissociate cell division from nuclear division: in this mutant, at the restrictive temperature, multinucleated amoebae are formed. We know that they are not little plasmodia by sophisticated genetic tests and a very obvious phenomenon: plasmodia can fuse with their own kind yet they eat their very own amoebae – even the multinucleated ones of hts 23 (Burland *et al.*, 1980).

This brings us to a final comment and another essential difference between the two alternative states – both vegetative, by the way: they can recognise each other somehow (yet by genes other than the many fusion genes (*fus*) discussed in plasmodia), probably involving the membrane surface. In this respect, the identification of already-detected differences in the surface antigens of the two cell types by immunology and iodination, can be very revealing. In this work (Kuhn, 1980), a stage-specific antigen for a plasmodium or an amoebae has been detected (the former not being directed against the slime), which can now serve as a sensitive marker to distinguish between the two cell types. Furthermore, the fusion gene products (once revealed), since they are expressed in plasmodia and not in amoebae (Carlile, 1976), will serve to discriminate the two cell types. Eventually, however, one must find out when the *fus* genes are first expressed, because that does help to stabilise a major developmental event: the irreversible amoeba–plasmodium transition (apt). However, we shall first look at the equally exciting, yet almost instantaneously reversible, amoeba–flagellate transition.

Amoeba–flagellate transition

This process involves a complex structural change of the amoebae and allows a different motile behaviour, utilising two flagella : a long straight one and a shorter bent one. Even when creeping on the substratum, the elongated swarm cell moves about like a worm, with the cone, from which the flagella emerge, being the front. At any rate, the amoeba becomes a different cell type, displaying polarity while developing a new organelle (Fig. 57). On the other hand, the transformation is facultative and after extended periods of axenic growth the capacity is lost. Actually, the

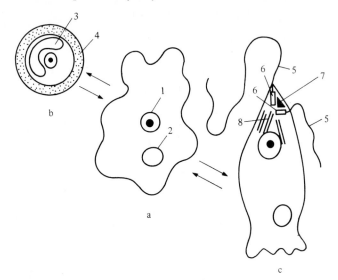

Fig. 57. Reversible amoebal differentiations. An amoeba (a) with a single nucleus (1) and contractile vacuole (2) can transform into a microcyst (b) with a large vacuole (3) and a solid wall (4), or a flagellate (c) with two flagella (5), two kinetosomes (6), one parakinetosomal body (7) and various microtubular arrays (8).

tendency to flagellate is quite differently distributed among the various myxamoebal species (Ross, 1957). It is not clear from inhibitor experiments with proflavin (a more efficient drug than actinomycin D in *Physarum*) whether RNA transcription is required, as only a delay in flagellation is observed. However, cycloheximide does inhibit the formation of the new cell shape, in particular the cone, and can even block the full outgrowth of the flagella after their formation has begun (Kerr, 1972). It is uncertain whether the assembly process of tubulin into microtubules is inhibited or the synthesis of tubulin itself. It seems that the building blocks (the tubulins, of which there are two that are quite different from those in other higher organisms) are already present because a pre-incubation of amoebae at 4 °C before inducing flagellate formation has no effect on it. However, it seems clear that this process requires energy, as dinitrophenol (DNP) is a potent inhibitor.

The flagella, the basal bodies and the nucleus are all firmly interconnected as a complex, which has been isolated and carefully analysed by serial sectioning and examination in the electron microscope (Wright, Mir & Moisand, 1979). Thus, it was found that this complex is highly structured and specifically organised. It can serve as an excellent example of intra-cellular morphogenesis. The two typical kinetosomes, one anterior and one posterior, from which the two flagella arise, are held together by several microtubules. Between them there is a fuzzy structure, the microtubule organising centre (MTOC). Beside the kinetosomes is a parakinetosomal body, devoid of microtubules. This asymmetry is also borne out by five arrays of microtubules, three originating from the anterior kinetosome and two from the posterior one. Where possible to count the microtubules, a fixed number is found, typically 35 for array 1 that connects with the nuclear membrane, and 39 for array 2. An interesting observation is made after application of cytochalasin A. It prevents the appearance of microtubules in the cone and also prevents the formation of the flagella (although the rest of the complex appears normally, in timing as well as in arrangement).

So far, the temporary transition of *Physarum* amoebae looks quite similar to a self-assembly process, which does not seem to require many new building blocks to be synthesised. However, as in formally similar events such as the morphogenesis of phage T_4, quite a number of genes seems to be involved in this morphogenetic process. The evidence stems from an analysis of a set of mutants that were isolated after mutagenesis of amoebae with nitrosoguanidine; these were unable to grow at 31 °C, yet appeared normal at 22 °C (Del Castillo, Oustrin & Wright, 1978). Surprisingly, 25% of these ts mutants turned out to be also defective in the amoeba–flagellate transition (fla⁻). By setting up crosses between these mutants, several *fla⁻* genes have been defined, and it could be shown that they were not just an unspecific consequence of some growth defect, as 6 out of 17 mutants could not flagellate, even at 22 °C (where growth was almost normal). Furthermore, these mutants (fla⁻) could be dissociated from the ts growth deficiency in a cross with a wild-type amoeba (*ts⁻, fla⁻* × *ts⁺, fla⁺*). Both factors freely recombined, yielding also *ts⁺, fla⁻*, as well as *ts⁻, fla⁺* progeny.

One mutant (TO 79) has been particularly revealing. In this case the amoebae became polarised as a cone formed, but no flagella were seen. In the electron microscope it was found that the kinetosomes had remained close to the nucleus (as in amoebae) and microtubules were completely absent. Instead, a large mass of microfilaments filled out and probably stabilised the cone structure at the front of the unflagellated 'flagellate'.

With the analysis of further mutants it may become possible to dissect this morphogenetic process into even more steps, and it will be rewarding to attempt 'saturation' with mutants for this event and find out how many different genes are involved in this non-essential, yet very delicate, structural transformation of myxamoebae (see Fig. 57).

Microcyst formation

Under unfavourable environmental conditions myxamoebae can round up, become immobile, and produce a wall. They become dormant and resistant, provided the conditions that induce this transition prevail; if the conditions change for the 'better', amoebae quickly excyst again (Gorman & Wilkins, 1980 for review).

In the laboratory, amoebae grown together with bacteria reach stationary growth phase when the food sources are used up. Then cysts form after a period of two to three weeks. This process occurs faster and more synchronously after the addition of mannitol (0.5 M), within a few days, and just within 24 hours in 0.2 M sucrose. It is assumed that these substances do not just change the osmotic pressure in the cytoplasm but somehow interfere with food intake by phagocytosis and thus indirectly cause starvation; just as an over-abundance of external water induces encystment, which, after all, is a dehydration process. Rapid encystment can also be quickly induced by a non-nutrient salt medium, if the amoebae have been grown axenically.

Morphological changes include a decrease in food vacuoles, and the appearance of vacuoles that contain fibrous material which is probably secreted as the extracellular membrane is built. The microcyst wall contains the same galactosamine polymer as the macrocysts, although it is much thinner. Other vacuoles appear and fuse to a single, large, central vacuole, which collapses as its contents are emptied out, and the whole cyst becomes flattened down (Goodman, 1972; Turner & Johnson, 1975).

The rough endoplasmic reticulum, though present as the wall material piles up, becomes generally reduced. The nuclear chromatin does not become condensed whereas the nucleolus does and looks inactive. This is consistent with the decrease in RNA and, probably as a consequence thereof, protein synthesis. The RNA becomes degraded and glycogen too, as encystment proceeds. However, neither an increase in proteases nor an extensive protein degradation occurs (in contrast to macrocyst formation). Furthermore, protein synthesis is required, as cycloheximide inhibits cyst formation. The addition of amino acids to amoebae inhibits encystment, whereas glucose, the most efficient carbon source for growth, cannot prevent the encystment. As in macrocyst formation, glucose cannot be taken up and metabolised during encystment. However, in amoebae external galactose is readily taken up and appears in the cell wall polymer.

From DNA measurements, which detected 0.3 pg DNA per amoeba (for 1C-content), it can be assumed that the microcysts are arrested in 'G_1 phase'. This raises once again the question of cell-cycle control in development and the role of a G_1 phase or, more generally, the cell-cycle concept. We have discussed earlier the free-wheeling mitotic cycle of the multinuclear plasmodium during balanced growth. Here, we can add, after surveying many unicellular organisms, that many of them have only a brief G_1 phase, if any, while they grow fast. However, an extended G_1 phase can appear when they grow slowly (Prescott, 1976, for review). In the particularly well-analysed case of *Amoeba proteus*, an arrest in G_2 phase has been documented, just as in the spores of *Physarum* (Ron & Prescott, 1969). If we remember that the cleavage cell cycle in embryos has neither G_1 nor a G_2 phase we may gain in understanding if we could eliminate these gaps from the cell cycle, conceptionally. Just that is happening in recent studies of cell-cycle control in mammalian tissue culture cells (Liskay & Prescott, 1978). These systems do not equal the situation in the living animal, having become adapted and often aneuploid in the ill-defined process of 'establishing' cell lines that keep proliferating while primary cultures do not. A startling conclusion has been reached: G_1 phase may be the consequence of some unspecific somatic mutation. The evidence comes from a Chinese hamster cell line that has no G_1 phase; it has been classified G_1^-. If it is fused with a subline having a G_1 phase (i.e. G_1^+) the hybrid cell has no G_1 phase. The crucial experiment is quite simple and convincing: if cells from two lines, which have a distinct G_1 phase, are fused, the G_1 phase is absent in the cell hybrids formed from them; in other words, a G_1^+ state has become G_1^-. One can argue that a complementation of two different growth defects (G_1^+a and G_1^+b) has taken place, restoring the G_1^- state, which may be normal for proliferative tissue cells, just as for protozoa and *Physarum*. Consequently, wherever a G_1 phase is present, the cells are just not growing at optimal rate. Indeed, in a G_1^- tissue culture line a G_1 phase has been created by lengthening the generation time in poor medium or by slowing DNA replication with a little hydroxyurea. Lengthening of generation time can expand the G_2 phase as well and one prediction has it that both gaps (G_1 and G_2) can be eliminated if the cell has all its needs catered for in excess. This is what has been achieved in germ cells in the organism, but not yet in the laboratory with tissue cells derived from them.

This raises a fundamental criticism of the cell-cycle concept and much of the work on proliferation control utilising tissue culture cells, and may indicate that some facet of cancer research has gone in the wrong direction.

If one turns to the cell cycle or cell kinetics in the living animal, there we encounter another surprise. As cells in a multicellular organism do not grow exponentially, there is of course a G_1 phase. In artificial in-vitro systems the G_1 phase is very variable in duration, probably because of somatic mutation, chromosome imbalance, improper food supply or utilisation. Nevertheless, S phase, G_2 phase, and M phase are quite stable. However, in contrast to in-vitro systems, G_1 phase in the living animal varies just as little as the other phases (Schultze, Kellerer & Maurer, 1979).

Let us now turn to the important implication of the G_1 phase which is supposed to include the commitment of a cell to pass from one state of determination (or differentiation) to another one. It is well established that such cells are post-mitotic. We can formally say that an irreversible event must happen at some time, and if as a consequence of this the G_1 phase gets longer (or shorter), nothing will be gained in our understanding of the differentiation process if we just say it happens in G_1 phase. Actually, the important event probably does not even take place during that phase defined as G_1^+. However, if we come back once more to the quantal cell-cycle concept, it is a round of DNA replication and some mysterious 'reprogramming of the chromatin' which seems to do the trick. The decisive event occurs before mitosis, i.e. before the cell arrives at the post-mitotic G_1^+com ('committed') stage. This alternative view of the classical cell-cycle concept contains more than pure semantics; it may throw some new light not only on the mechanism of commitment in 'normal development', but to various derangements thereof that are lumped together as in the 'cancer state'. That state, which can be formalised as G_1^+can (for cancer) then turns out as a defect in differentiation and not in cell proliferation control.

What does this have to do with *Physarum*? Three points can be made. First, this system was set up right in the beginning by Rusch as an inexpensive model to study the switch from growth to differentiation, approaching the cancer problem through basic research, in an attempt to find some substance that promotes differentiation and thereby inhibits proliferation. Second, the G_1-phase arrest, in *Physarum* cysts as in other protozoan cysts and any 'unhappy' cell in culture, may just be a state of partial metabolic inactivity, yet readily reversible. However, the tight coupling of mitosis and initiation of S phase in *Physarum* over very extended cell cycles, from 8–36 h in the plasmodium or from 4–48 h in the amoebae, as opposed to the putative G_1 phase in the cyst, might tell us something. Therefore, it will be imperative to measure DNA contents in nuclei of cysts that are induced by many different means, in particular those that make cysts within hours, and a close analysis of some 16 000 metres of film on the life history of *Physarum*, available at the Film Institute in Göttingen, Germany, may be very rewarding in this respect. A third point can be made with respect to (*a*) gap-less embryonic cell cycles, (*b*) G_1^- proliferative cell cycles in various organisms, and (*c*) the post-mitotic state of most, yet pre-mitotic state of some, differentiated cell types. What if these states, like in established cell lines *in vitro*, are due to somatic mutations, which occur along developmental pathways, where alternate decisions are required successively and lead each time to a new state of commitment, which is so stable that many have wondered how it gets transmitted over hundreds of cell generations, like a mutation? Well, they may be 'mutations'.

At any rate, the formation of microcysts is readily reversible by a process, called excystment, which is initiated by any 'favourable' condition, including a supply of a little (not too much) water, yet inhibited by too much food (such as 0.1 M sucrose), disturbing unspecifically the osmotic balance in the cytoplasm.

Encystment involves a partial lysis of the wall, and the amoeba crawls out once again as a mobile cell through that little hole, the ostiole. Protein synthesis is required, as cycloheximide inhibits this process. Gene expression is also activated, as deduced from the reappearance of the rough endoplasmic reticulum. It is not clear from inhibitor experiments whether new RNA synthesis is also essential.

Here, genetic techniques can be used to determine which of the new proteins are specific for encystment and which are required to reactivate the metabolism and general growth processes. A very simple and efficient selection technique has been devised for amoebae that cannot excyst at a restrictive temperature because of a mutation. While all other individuals will do so, and get eliminated by treatment with Triton X-100 through cell lysis, the one that cannot excyst remains alive and will become an amoeba, when the lysed cells have been washed off with fresh medium at the permissive temperature. As it multiplies, enough amoebae will be formed as a clone. Many of these are being tested and analysed genetically. However, biochemical analysis is also possible in a search for what goes wrong at the restrictive temperature.

As we look over macrocyst and microcyst formation in *Physarum*, we observe many similarities and some differences in these differentiation processes, which also resemble certain developmental features during the encystment of small soil amoebae (like *Acanthamoeba*, a diploid cell) or the dormant state of the haploid social amoebae of *Dictyostelium* or its relative *Polysphondylium*.

We can ask whether this dormant state, which is a facultative side-line in the life cycle of *Physarum*, is good for something. Microcysts are potentially useful for the survival of the species, when a sudden change occurs in the environment making it impossible for the diploid plasmodium (or its macrocysts) to continue growth (for discussion see Collins, 1979). For example, in the laboratory, if one throws in cycloheximide, one will kill all plasmodia in that culture dish right away, as it is unlikely that some 10 000 small macroplasmodia are living side by side. On the other hand, 10 000 cysts or amoebae take up little space and are only a fraction of the potential progeny of one fruiting body. They will contain some resistant mutants that go on growing in the presence of the drug. Generally, such genes, favouring survival, can be immediately effective in the haplophase, as they are not cancelled out by the other wild-type allele in the diploid plasmodium. Whether this mechanism plays any role in the natural environment is not known. Under favourable and adverse growth conditions alike, microcysts are eaten up by a vigorous plasmodium, even more often than the seemingly unprotected amoebae, as has been demonstrated in a recent film.

8

Becoming a higher organism:
the amoeba–plasmodium transition

The heading of this chapter implies that an amoeba can turn into a plasmodium. That is wrong; it has first to become a different creature, a gamete. Although it looks very much like that simple amoeba which divides, makes flagella, or encysts, it must be 'competent' to become a plasmodium. We encounter the same problem of conditioning that we discussed in sporulation, the other irreversible pathway in *Physarum* development.

Just as mutations, the driving force of evolution, happen at random, thus making biological development a historical event where the future direction is not predictable, to discover and make developmental mutants in *Physarum* adds a biographical dimension to our search to explain the developmental process. This eliminates, for a while at least, the application of logic and the principle of cause and effect as reliable tools in the analysis of any developmental system, and creates utter confusion, as ever more mutants become known.

On the other hand, we stand at the threshold of multicellularity, that great step in evolution which enabled tiny or 'lower' eukaryotes to become large and diversified 'higher' ones. Nothing is known about the principles involved in this transition, which repeats itself with stunning success when an egg cell continues its life by cleaving into many cells before growth and differentiation set in. The higher the organisms are evolved, the more they depend on this susceptible unicellular period in their life cycle. A 'lower' animal, say a worm, or many plants can be cut into pieces, and each piece can regenerate, just as the breaking up of the plasmodium into macrocysts in *Physarum* creates more plasmodia.

The fascinating thing about *Physarum* is that the whole process can be analysed and dissected as several pathways or steps, and although more and more mutants are described, the genes involved are really very few, maybe just one or two, albeit complex, loci. What a prospect, where millions of individuals can be handled and experimented on in the single cell stage (as amoebae, comparable to a permanent clone of germ cells), and each one can be analysed in the ensuing macroscopic life forms: the vegetative multinuclear plasmodium and the cellular sporangium. These features, and the uncoupling of meiosis and sexual fusion from the establishment of multicellularity, are unique in *Physarum*. Even if they were not, their analysis must lead deep into the unknown, even unimaginable mysteries of life.

Dissociation of cell fusion from cell transformation

Let us recall the flexibility of forming a plasmodium from amoebae (Fig. 13), and then make use of it. First, it has been shown that each isolate collected in the field has a different mating-type system (14 at the last count). In any one isolate there are only two alleles of the mating type (*mt*) expressed, say + or −. However, in all combinations of amoebae from different isolates, each *mt* pair is found to be different yet can be assigned to the same *mt* locus, making it a multiallelic complex locus. As long as any two amoebae differ in the expression of *mt* (symbolised *mt* $\neq$) they can cross and develop into a diploid plasmodium. These amoebae are homotypic, incompatible for sexual mating (fusion of gametes), i.e. they are heterothallic.

Some strains of *Physarum*, among them the Colonia strain in particular, are different. Plasmodia arise without sexual fusion, i.e. asexually. Hence, such strains are selfers. Furthermore, they are haploid throughout the life cycle. They have a special *mt*, yet it is also located at the same complex locus. The amoebae of the Colonia strains actually have the option either to self or to cross with other (heterothallic) amoebae, and the experimenter can help that crossing prevail, since selfing can be blocked at high temperatures. This already shows that the *mt* gene may not be involved in cell fusion at all.

A different gene has been identified that controls fusion of amoebae: it has been called *rac* (for rapid crossing) or *mat B* in two different laboratories, where it has been independently discovered (Dee, 1978; Palotta *et al.*, 1979). Once again the two amoebae must have different alleles (symbolised as *mat B* $\neq$) to fuse with each other. And again, a multipolar system is at work, as several alleles of the gene *B* have been detected. This then makes sexual mating and plasmodium formation an at least tetrapolar system, a composite of two systems, one controlling mating (the multiallelic gene *mat B*), the other plasmodium development (the multiallelic gene complex, *mt A*). From now on we shall refer to gene *B*, for mating, and gene *A* for initiation of plasmodium development.

Two systems can be taken apart: when we put amoebae together that harbour a different gene *B* allele (*B* $\neq$) but the same gene *A* allele (*A* =), they fuse but they do not make a plasmodium. A film analysis has clearly shown (Holt & Hüttermann, 1979) that about two hours after fusion both nuclei start prophase and an open mitosis takes place, yielding two diploid nuclei. Then the cell divides by cytokinesis, thus producing two diploid amoebae which can go on dividing (Fig. 58*a*).

The sequence looks different when two proper amoebae fuse (they are compatible as they are *A* $\neq$, *B* $\neq$). Here, after cell fusion, the two nuclei fuse

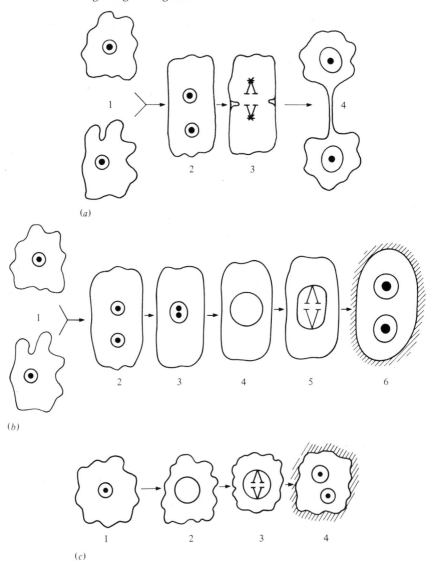

Fig. 58. Plasmodium initiation involves two steps. (*a*) Two haploid amoebae (1) with different alleles of the fusion gene, *mat B* (*B*≠), yet identical developmental gene *A*, *mat A* (*A*=), fuse and result in a binucleate cell (2); chromosomes of both nuclei enter a common spindle (3), and cell divisions result in two diploid amoebae (4). (*b*) Two haploid amoebae (1) with different alleles at *mat A* and *mat B* (*A*≠, *B*≠) fuse (2); nuclei fuse (synkaryon) (3), and the fused nucleus swells (zygote) (4); intranuclear mitosis (5) leads to diploid plasmodium (6). (*c*) One amoeba of the selfing strain (1) reveals nuclear swelling (zygote equivalent?) (2), followed by intranuclear mitosis (3) and transformation into a plasmodium (4).

in interphase, forming a zygote nucleus which shows two nucleoli. After about two more hours a closed mitosis occurs, yielding two diploid nuclei, in which a single nucleolus becomes reconstructed. Cell division is left out (occasionally it does occur in the two-nucleus state but not at higher nuclear numbers) (Fig. 58b).

No one would expect much if two fusion-incompetent amoebae meet (being $B =$ and $A =$); nothing should happen, as they cannot fuse. Nevertheless, fusion does occur, yet very rarely – at least 1000 times rarer than in the $B \neq$ system. This can be easily scored on a plate of millions of amoebae, and explains why the B gene has remained so long undetected.

One can speculate that the multipolar *mat B* system is labile, as occasionally a B_1 becomes a B_2 or vice versa. There is a situation in yeast, which is quite similar to this unstable mating behaviour. There it can occur as often as once per cell cycle, i.e. the mating type can be switched around since one individual can act as a male in one generation, and as a female in the next generation. As an explanation for the frequent yeast *mt* switch, the cassette hypothesis has it that the *mt* gene is a composite of two pieces of DNA, where one piece gets taken out of a 'cassette' of similar gene pieces and transposed to and inserted into the other piece, which serves as a 'cassette deck'. Direct evidence for this hypothesis is accumulating from recombinant DNA analysis (Nasmyth & Tatchell 1980; Strathern *et al.*, 1980).

As another permutation of plasmodium formation, diploid amoebae can be constructed (by the process shown in Fig. 58b), which harbour different alleles A and B ($A \neq$, $B \neq$). They would not have to fuse to turn into a plasmodium – and they do not; they act like 'selfers'. However, their progeny become haploid and cannot self anymore. This result tells us that it takes two steps for plasmodium development; one leads to the initiation of a plasmodium as long as the A alleles differ ($A \neq$). Yet something else must happen to maintain this state for the following generations. That step does not seem to work in these diploid amoebae.

As the last case, how do the real selfers which are haploid amoebae (not diploid amoebae as just described, which are heteroallelic to begin with, i.e. $A \neq$, $B \neq$), turn directly into a plasmodium? Again, film analyses have been made, but little can be seen aside from the decisive act: the occurrence of a closed instead of an open mitosis. However, in each case an enlargement of the nucleus takes place (at the time of the zygote formation in sexual crossing) before the first plasmodial mitosis occurs (Fig. 58c). This 'pregnancy' is probably not associated with additional DNA synthesis. It is suspected that it indicates an active phase of RNA transcription.

It will be most rewarding to study cell-cycle phases (and events) associated with the initiation of plasmodium formation. Just to remind you, plasmodia, once established, recognise each other and can fuse if they have an identical set of fusion genes. Hence in plasmodia we encounter, in contrast to the situation in gametes, a heterotypic incompatibility system which can be cleverly employed.

This discussion of plasmodium formation, although already the second step in the plasmodial developmental pathway, was necessary. Now we have an idea of what to look for while the essential first step, gamete formation, takes place.

Regulation of gamete formation and plasmodium initiation

As we have just said, gametes look just like amoebae, but they are competent to fuse sexually (or not to fuse) and to create plasmodia. As in the process of gaining competence to sporulate, we need to rely on indirect parameters: one is the growth state of a population of amoebae. Thus, about 5×10^5 individuals must crowd together, before gametes are detected, no matter whether one starts out with a hundred amoebae (and waits for weeks) or with 100 000 (and waits for just a couple of days) on the culture plate. This group effect smacks of factors that are produced by the amoebae and act 'positively', i.e. they encourage the transition of an amoeba into a gamete.

Indeed, a soluble diffusible factor has been described which enhances plasmodium formation. The material is labile and has not been further characterised (Youngman *et al.*, 1977). This factor is not some sort of sex hormone, determining whether an amoeba becomes a male or a female gamete (i.e. *B*-gene function). This factor influences plasmodium development (i.e. *A*-gene function). That can be best seen if a large number of selfing amoebae is put on one side of a filter. This set-up promotes development of plasmodia in a pair of heterothallic clones of amoebae on the other side of the filter. Therefore, the factor can substitute for the group effect which is required in normal development. This has been used in filming the plasmodial initiation and thereby saving a lot of film material.

This factor could be a positive acting substance, i.e. a new gene product, triggered late in exponential growth phase; it could also be a product that is constantly secreted and reaches threshold only in a dense population. Of course, the factor could also be a negative-acting substance, like one of the secreted proteases, that needs to be eliminated (remember, only well-fed amoebae secrete proteases).

It so happens that the addition of proteases (subtilisin and α-chymotrypsin), while having no effect on the growth and proliferation of amoebae, can block the formation of plasmodia altogether, as shown in the following experiment: two clones of consenting amoebae (i.e. competent amoebae that become gametes of the state $A \neq /B \neq$), in the presence of protease from 1 h before mixing until 2 h after mixing the two clones, will not form any plasmodia. This tells us, once again, that the outer membrane (the target of the enzyme) is specifically involved right at the time of plasmodium initiation.

Surprisingly, only a few plasmodia are ever seen if amoebae get competent in a crowd, just one from at least a thousand complacent partners. Actually, this is the optimum result of plasmodium formation. As shown in Fig. 59*a*, once the lag phase is over, plasmodia occur at an increasing rate. However, the more plasmodia arise, the fewer will arise from the remaining amoebae. That must deter the molecular biologist who wants one test tube with amoebae, one with gametes, and yet another one with plasmodia, and not such a mixture. Well, he can have that, too. All he needs to do is take a temperature sensitive (ts) *mat B* mutant strain, that cannot make plasmodia at high temperatures. Most amoebae in a mixture, and virtually all of them in a selfing strain, seem to become gametes in the heat, and turn into plasmodia in the cool, with good synchrony. This is the system of choice (Fig. 59*a*) and somebody is going to exploit it. (To be fair, one must add that this complete conversion of many gametes into many plasmodia does not yet happen in liquid medium, but only in the presence of killed bacteria on an agar substrate.)

Such observation holds the key to the paradoxical phenomenon that the more amoebae you take, the fewer plasmodia you get: a plasmodium once formed releases its own factor, which inhibits plasmodium formation. Of course, one suspects the slime as being responsible, particularly as a slime-halo is already seen in the film of a plasmodium with just two nuclei, i.e. one that is just a few hours old. This also reminds us of the slime component that has been shown to block amoebal division (maybe cytokinesis of the tiny plasmodium as well?) in very low concentrations, reminiscent of some sort of chalone.

In summary (Fig. 59*b*) we now have some evidence for a possible soluble factor (even if it be the lack of a protease) inducing plasmodium formation, and a putative inhibitor produced by plasmodia repressing plasmodium formation.

But there is yet another positive effect brought about by the crowding of amoebae. This is the crucial test: take amoebal clones from each of two heterothallic plasmodia (gene *Ax* and gene *Ay*). When they are mixed together they make some plasmodia (as *A* ≠). Now mix in a third clone of an asexual strain (say gene *Az*) of a kind that cannot undergo sexual fusion (i.e. a mandatory selfer, as shown by clear-cut genetic tests). Such trisomes reveal an unexpected result: many of the haploid heterothallic amoebae (either *Ax* or *Ay*) now prefer to self instead of mate with each other. Therefore, instead of diploid plasmodia (expected from heterothallic mating) haploid selfed plasmodia arise. Since chelating agents such as EDTA inhibit this effect of the third party, which by design is already asexual (i.e. specialised to turn a gamete into a plasmodium without sexual fusion), cell contact must be the way 'selfing' in the heterothallic gametes is brought about. Once again, we find the membrane involved in plasmodium initiation. The preliminary search for different surface antigens between amoebae and

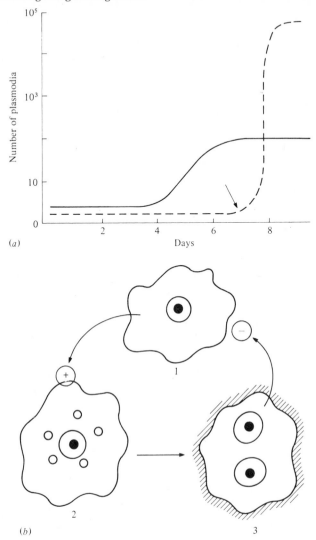

(a)

(b)

Fig. 59. Mechanisms of amoeba–plasmodium transition. (a) Kinetics of plasmodium formation in a large population (10^5) of amoebae. After a lag-period, plasmodia form first at accelerating and then decreasing frequency, leaving many individuals in the amoebal state (solid line). In a ts mutant of plasmodial formation (dashed line) no plasmodia appear at the restrictive temperature (30 °C), yet all amoebae become swiftly transformed into plasmodia at the permissive temperature of 22 °C (arrow). (b) Putative regulation of plasmodium formation during lag-phase. Amoebae (1) produce a positive acting factor (+) which allows some to become gametes (2) (amoeba–gamete transition). These are assumed to express gene *mat B* (o) and initiate a plasmodium (3) (gamete–plasmodium transition), which in turn secretes an inhibitory factor (−) that prevents formation of further gametes.

gametes has not been successful, although plasmodia and amoebae do differ immunologically (Kuhn, 1980). Much has still to be learned about the surface of the gamete membrane, and how gametes fuse (if they fuse). Although the gamete is covered by a glycocalyx, so far no lectins have been detected, which could specifically hold the two partners together. It is completely unknown how the gametes come close enough (say within 1 nm range) to initiate cell fusion. On the other hand, plasmodia with their massive slime coats fuse alright, and very fast.

While the above observations stress cell contact in initiation of plasmodium formation, the ionic conditions of the environment also play a role (Shinnick *et al.*, 1978). Thus, mating does not occur if the optimum pH 5 is raised to pH 6. However, this barrier can be overcome at high ionic strength or in a mutant, appropriately called 'imz' (for ionic modulation of zygote formation).

There is one final comment to make concerning the state of 'competence' which we have found to be linked to the growth state (cell-cycle state?), exogenous factor(s) and cell contact. The question arises whether the gamete may not just be competent to fuse and/or initiate plasmodium formation. It may be a different cell type altogether. Let us illustrate this with a brief look at a sperm cell in an animal. It is haploid, very complex in structure (some say the most intricate cytodifferentiation known), and very short-lived. A sperm is highly specialised in structure and function and destined to die if it does not function fast. In our case the question is: is the gamete of *Physarum*, which is highly specialised in function but not with respect to structure, committed? Or, in other words, has the gamete experienced 'commitment' without differentiation, and most importantly once committed is the gamete destined to die as it cannot turn around and become a naïve amoeba anymore? If it does change into an amoeba, amoeba–gamete transformation is reversible, just like cyst formation. If it does not, we have to add yet another irreversible directed section in *Physarum*'s life cycle – short but essential. This leaves us with a problem for the selfer. When shall we define it to be no longer a gamete and already a plasmodium? Maybe there is not a problem once we learn that competence and commitment belong to the developmental pathway of the gamete, and differentiation to the functioning of the *A* gene, i.e. the pathway of the plasmodium.

Several attempts are being made to find out what really changes after the mating, or after the gamete–plasmodium transition if there is no mating, when a microscopic gamete turns into a macroscopic creature. The second case (the selfer) is the more promising system to study since the function of the mating gene (gene *B*≠) is not required. Even then things are quite variable when one looks over the behaviour of various selfers: in most cases a plasmodium has a closed mitosis and many nuclei. It fuses with a large plasmodium, yet eats amoebae or cysts. Rarely, even at the one-nucleus stage of a selfer, a 'preplasmodium' is formed, as it fuses with a plasmodium yet devours amoebae. Also, one cell division can occur after the first open mitosis. Nuclear and membrane events seem not always to occur in a fixed order. Now that an antibody against slime, and two more, one specific for either amoebae or plasmodia, are available (Kuhn, 1980), it can be asked when these specific markers first appear.

It is already clear, however, that even without sexual fusion (i.e. in the selfers), the surface does change at plasmodium formation. It is such an early event that one can speculate that the surface changes are really the cause of the transformation of the gametes. To follow this argument, the fusion behaviour has been analysed further. At least three sets of fusion genes are expressed in plasmodia (and they must completely match if two plasmodia are to fuse, see above). In the amoebal state these genes seem to be repressed as amoebae do not fuse, even if the fusion genes are all the same. (Gametes must be different at the gene B to mate. This situation defines two different kinds of fusion, one sexual the other somatic, even if fusion happens in the one-nucleus state. The gene B seems to be repressed in plasmodia.)

Now one can do an exciting experiment: take two different selfing clones ($A =$), one having *fus* A_1, the other *fus* A_2 (these alleles are codominant, i.e. when they are united in one cell; this plasmodium cannot fuse with any of the selfed plasmodia, as those are either *fus* A_1 or *fus* A_2). Both, *fus* A_1 + *fus* A_2, should not be together if two different selfing clones are mixed. The exciting thing is that one does find both A_1 and A_2 in one plasmodium in two situations: sometimes as heterokaryons (i.e. in haploid nuclei), sometimes as hybrids (i.e. in diploid nuclei). There are still two ways to explain this unexpected result: (1) the heterokaryon can arise either by a mass fusion of haploid amoebae (maybe something went wrong with the sexual fusion loci B), and then the whole heterokaryon turns into a plasmodium, switching on both the *fus* A_1 and A_2 genes; or (2) 'preplasmodia', or young plasmodia, can fuse with each other, irrespective of their fusion genes, because the *fus* genes at this early state of the plasmodial life are not switched on yet. That is an essential distinction, if we wish to be sure that plasmodia are plasmodia because *fus* A genes function or the other way round: i.e. fusion gene expression is a consequence of determination of the plasmodial state.

But how do the diploid hybrid plasmodia (*fus* A_1/A_2) arise? They should not exist, since the two strains used have identical alleles of A ($A =$). (Remember, if the alleles of the differentiation gene A are the same, no plasmodia should form, instead diploid amoebae should arise). Well, such diploid plasmodia are very rare, and we are left with the speculation that a mutant at the developmental gene A locus has occurred, so that the two are no longer alike. (Of course, it cannot be excluded that in this case we have an example of homothallism, after all.)

This speculation brings us a long way when we consider such mutations in a large plasmodium (say a haploid selfer like Colonia), where millions of nuclei are shuttled around and some are always killed off anyway (at a steady proportion of 0.5%). These may all be somatic mutations, that do not injure the plasmodium. But such a mutation could help in survival: if the postulated mutation occurs in gene A, there would be one nucleus that can now fuse with any other haploid nucleus of the plasmodium and become diploid. This describes the well-known phenomenon of parasexuality that many microbiologists have to contend with in their genetic studies, including some of the highly developed genetics in *Dictyostelium* (Williams, Kessin & Newell, 1974; Loomis, 1975). In *Physarum* this fused nucleus can proliferate and produce many diploid nuclei. When the time comes to make sporangia in that particular plasmodium, the haploid nuclei fight their abortive meiotic attempts, and some manage to come through unharmed, making a 1n spore that can give rise to a new selfed plasmodium. However, the putative parasexually derived diploid nuclei have no problem with meiosis and end up in mature haploid spores more frequently. Now imagine that the selfing Colonia strain is defective in some way (it cannot be grown as balanced plasmodia, no matter how well you feed it; Turnock, 1980). This limitation has a genetic barrier as its progeny are just as weak. Therefore, adverse conditions, like heating up the whole plasmodium at 30 °C may really hurt the sick nuclei, and only the diploid nuclei give rise to viable haploid spores. This may sound very speculative, but this is a puzzling result that has been observed in the Colonia strain (being the source of most genetic work in *Physarum* as it is haploid), where a heat shock has increased the viability of spores from 2% to 80% (Laffler & Dove, 1977).

This episode, together with the foregoing discussion of the establishment of the plasmodial state, reminds us of the similar situation surrounding the occurrence of meiosis (usually after but also before sporangia have formed) or the general cell-cycle concept: timing of essential events is quite flexible. Consequently, it is the events that count, making it impossible to set up a unique temporal sequence for developmental pathways, and making conclusions from temporal sequence alone meaningless.

This brings us to the main concern of *Physarum* development and a major question of developmental genetics in general: what are the essential events in gamete–plasmodium transition?

If that event were the establishment of a closed mitosis from an open one, this would be quite interesting for an analysis of a mitotic peculiarity, but would make *Physarum* quite obsolete as a 'model system' of developmental biology. However, we have just detailed that a plasmodium is a plasmodium already at the uninucleate state, and that the nucleus as well as the membrane are altered. Mind you, these membrane changes have nothing to with sexual mating (the $B\neq$ situation) or fertilisation in other organisms. If we cast a last glance at sexual mating, we are reminded that the nuclei fuse during interphase, and only thereafter do they become different. In this case, involving sexual mating, they become diploid. But in the selfing case they get a little bigger at that time, too. That is the moment of truth: more than a unicellular organism is created (the gamete has become 'pregnant'). This seems similar to those higher organisms that can develop by partheno-genesis. However, the situation in *Physarum* is much simpler, as development is dissociated from meiosis and oogenesis. We suspect that the great simplification has become possible because as soon as the gamete has attained its supra-unicellular state it does nothing else but grow in balance, displaying free-wheeling cell cycles, shuttle, migrate, and monitor the out-side chemotactically. The main conclusion is that a plasmodium does not differentiate at all, unless it gets very hungry, goes completely off balance, and is illuminated by the sunshine.

All other higher organisms or true multicellular metazoa have an additional important decision to make right at the beginning of their supra-unicellular state, which equals the zygote. If they are animals they must determine which cells carry on the gametes (the germ line) and which should make the disposable body (the soma); if they are plants they can leave this decision open for a while and settle for a shoot or a root. In my view this cannot take place without elaborate, complex, and, to be honest, mysterious processes that surround 'oogenesis'.

In *Physarum* then, we might encounter a unique situation, a clear-cut example of switching from a unicellular state to the next higher organisation of eukaryotic life. As it happens the result is a plasmodium, but in reality this may be a straight line to metazoan life, and, to the developmental biologists, more illuminating than the rest of the life cycle of *Physarum*, which may well be a cul-de-sac in an evolutionary sense, as has been said by Cummins & Rusch (1968).

So far this has only been realised by only a few geneticists, among those are Jennifer Dee in Leicester, Holt in Boston, and Dove in Madison, with their respective collaborators. Formal genetics of *Physarum* had a slow start but sprouted and has become quite confusing to a non-specialist. How-ever, as a naïve non-geneticist, looking from a safe distance, and after some contemplation, painful at times, it seems really quite simple. In other words, the system seems to be ready for a combined attack by all sorts of tools of molecular, cellular, and developmental biology. Let us now come to the genes involved.

Genetic analysis of plasmodium formation

Now that we know how a plasmodial state begins, let us look at it the formal geneticist's way: disturb it by mutagenesis, and assign 'genes'. With a number of tricks available a real orgy is going on (see Gorman & Wilkins, 1980, for an introduction).

There are two methods of approach. Either one begins with a selfer (an amoeba that turns into a plasmodium when grown as a clone) and searches for mutants that cannot self. Such an individual is now unable to make plasmodia if grown as a clone (although proliferation as amoebae is not disturbed). Or one sets out with an amoeba that cannot make plasmodia in a clone (like all 'normal' amoebae from heterothallic strains) and searches for those individuals that can do so. In both cases many controls are necessary to make sure that one really has a mutant at hand. This is done by crossing with cleverly constructed tester strains and analysing the progeny, i.e. asking how the amoebae behave in the following generation, or how stable the mutant is.

Several research groups have started out with the haploid Colonia strain, since a defect in a differentiation gene would show up in the haploid plasmodial state, and all have detected many mutants that could no longer self. The mutant phenotype looks just like most *Physarum* strains, i.e. it is heterothallic. The mutant amoebae (let us call them dif⁻) need a partner to recombine the mutant *dif⁻* gene with the *dif⁺* gene from the wild type. If that is done by providing the appropriate partner (a tester strain that differs at gene *B*, so gametes fuse, and contain an allele of gene *A* that is different from the original strain to assure the $A \neq$ situation), a diploid plasmodium arises. Two questions are then put to its haploid progeny amoebae: (1) can you self (are you still *dif⁻*?), and (2) what kind of gene *A* do you have – if any? In most cases the answer has been that the progeny amoebae do not self (they are stable *dif⁻*), but none of them have taken over the gene *A* allele from the tester strain. That means that the genes *dif⁻* and that particular *A* allele (from the tester strain) do not recombine easy. Therefore, they must be on the same chromosome. In some very extensive tests, they have not recombined in 10 000 crosses. It seems, that both genes are so close together that they are part of the same gene locus, actually they may be allelic to each other (hence: multi-allelic *mat A*-complex locus). In other words, many different people all over the world, using many ways of mutagenising amoebae and searching for the *dif* gene phenotype, come up with the same mutant, i.e. with the same gene. Of course, each mutant has been given its very own name, but they may still be the same mutants. Some of these are temperature-sensitive, i.e. they do allow differentiation at the permissive temperature. At the restrictive temperature, even already-initiated plasmodia become blocked. That is a very important result, as it tells us that the *dif* gene must be continually switched on to sustain the plasmodial state and not just for a short while in the mononuclear state. Temperature sensitivity also indicates that a gene product (a protein) has to be made that is responsible for the temperature effect.

The next question then becomes: how do such individual mutant amoebae behave when they are allowed to mate? If the cross is set up so that gene B is different ($B \neq$), do the two mutants ($dif_1^- \times dif_2^-$) help each other to become a plasmodium (i.e. develop the dif$^+$-phenotype), or have they become only diploid amoebae (dif$^-$ phenotype)? If they turn into a plasmodium they complement one another because they are really mutants of different genes, each making a different product, which can act in concert with the other. If they do not complement, they might well have the same problem: they are deficient in the same function, i.e. in the same gene. One then arranges all mutant strains that complement into a complementation group. Doing that reduces the many dif$^-$ mutants to four groups. Maybe just four different functions can be found that can disturb plasmodium formation. The situation is even more simple as two of these groups contain more than 90% of all mutants (classified as dif A and dif B). Even more revealing is the fact that all mutants in these two groups are located on the same chromosome as gene A. Despite many attempts, neither of these two mutants could be separated from gene A by crossing. Consequently, they are all part of the same gene, they are also allelic.

This makes gene A, the turntable of initiation (and maintenance) of the plasmodial state and perhaps responsible for the zygote state, essential in multicellular life forms. However, *dif A*, *dif B*, and A are different genes as well. They come out of a cross unharmed and can be individually traced in the progeny amoebae, therefore this is a truly complex locus.

It also seems that this single locus can only change one way: either it allows selfing or it does not. It really looks like a switch that is 'on' or 'off'. This reveals a 'chicken-or-egg' situation. If you start with a selfer (being dif$^+$), any mutant in the *dif* gene becomes a non-selfer (i.e. dif$^-$, phenotypically a heterothallic); if you begin with the latter, you can only get the former.

If such a ping-pong mechanism exists, what happens if one takes a mutant clone (dif$^-$ like dif A) that has arisen from a selfer (dif$^+$), and now selects for that mutant which can self again (i.e. return to the original phenotype, dif$^+$)? That mutant should be a revertant of the particular dif A mutant to the corresponding dif A$^+$. Such a mutant has really been found, by design, and it must have happened in the laboratory unnoticed in the strain designated Cld (in that strain, d stands for delayed selfing of a subline of the Colonia strain). As it turns out, the strain Cld is really a mutant of the dif$^+$ gene, i.e. a dif$^-$ state that reverts frequently (i.e. leads again to the dif$^+$ state).

We recognise right away that the all-important gene A locus is quite unstable, and that may be a telling tale. But, to finish the first approach: i.e. start from a selfer (dif$^+$), look for mutants that cannot differentiate (dif$^-$), and revertants of dif$^-$ back to dif$^+$. The genes of the four complementation groups have been given different names by different people (*dif A* = *apt C* = *npf C*, closely linked to the switch gene A, and *dif B* = *apt B* = *npf B*, also linked to gene A; *dif* stands for differentiation defect, *npf* for non-plasmodium former, *apt* for deficient in amoeba–plasmodium transition) (Anderson, 1979).

This leaves two additional genes, *npf A* and *apt A$_1$*, which are not linked to each other nor to the main switch gene A. They have occurred rarely, the former in 2% of the mutants, the latter just once. Nevertheless, they tell us that differentiation of the plasmodium requires at least two genes other than the complex gene A locus.

Let us now turn to the alternative approach and search for selfers (i.e. dif$^+$) derived from heterothallic amoebae (by definition here considered dif$^-$). The result is surprising. In each and every case, if you wait patiently for say 2–3 weeks (until the clone has produced a respectable number of 10^{10} amoebae or has made plaques of 20 mm diameter), you will find a selfer. That should not be so; therefore this kind of plasmodium formation from one heterothallic amoebal clone has been called 'illegitimate'. Of course, one could argue that in such numbers there must have been a mutant that has become a selfer: but that is not the case. If you analyse genetically such illegitimate plasmodia by the amoebae they produce after sporulation, they have not changed their genotype. They cannot self just as their parents could not. They have maintained their specific mating type (say their gene A allele).

As this phenomenon is not due to a proper mutant, let us postpone further analysis until we have discussed some real mutants, which show up in a shorter period of time in a clone of heterothallic amoebae, when a plaque has not become larger than 1–2 mm in diameter. These are real mutants, i.e. their children remain selfers and they also retain their parental gene A allele. They were named 'cat' (for clonal amoeba–plasmodium transition) by Gorman *et al.* (1979). In absolute terms they are anything from frequent: 10^{-9}–10^{-8} in random culture, and maximally 10^{-6} after mutagenesis (as opposed to 10^{-5}–10^{-6} for random or 10^{-3}–10^{-4} for mutagenised structural genes, like that for thymidine kinase, TK).

But they occur frequently enough to ask the obvious question: as we know already that the parental gene A allele is still there, where are the cat mutations located? First of all it has become clear that the selfer phenotype is brought about by a single Mendelian factor, as a cross of a cat mutant with a tester strain that is cycloheximide resistant (*cat* × *cyc^R*) yields 50% of progeny that can self and are resistant (cat$^+$, *cyc^R*). On the other hand, mutants of the *cat* gene, in crosses with strains having the parental gene A allele, have almost never recombined with that gene. That means that the two genes (A and *cat*) are so closely linked together that, despite many crossings, they have not been separated as yet. This is quite similar to the previous situation, searching for selfers that cannot self any more, which led to the assignment of the *dif$^-$* gene somewhere inside the complex gene A locus.

Now we can put the *cat* gene in as well. However, as for the *dif* gene, there is one exceptional cat mutation that is not linked to the parental gene *A* allele. This indicates that at least one additional gene can be involved in becoming a selfer.

Using a different approach, a different class of very rare mutants has been discovered that also self (Truitt & Holt, 1979). In this case, the creature that crawls out of the spore immediately turns into a plasmodium; it looks like an amoeba but it is not. Unfortunately, only one spore out of about 10 000 can do that. However, this is a clear indication of genetic regulation at the other side of the life cycle of *Physarum*, where a plasmodium turns into spores. Consequently, these mutants are named: alc (for amoeba-less-cycle). This then marks an important developmental decision: plasmodium–amoeba transition, and one might speculate that the rare spore is really a uni-nucleated macrocyst that happened to end up in the wrong place, the sporangium. Fortunately (one is tempted to say, as otherwise the genetic analysis might get too simple), this alc mutant is unlinked to any of the relevant developmental genes, defining the beginning of the life cycle: the amoeba–plasmodium transition (apt). As a reminder, these include: *mat B* (the multi-allelic locus for sexual fusion, mating), the complex locus gene *A* (the developmental switch, including all the frequent mutations of dif A, dif B, and the cat mutants, and the rare mutations – or stable genes? – *npf A*, *apt A₁*, and *cat x*).

In summary, this search, starting out with heterothallic strains, has stressed the great significance of gene *A* functions in establishing the macroscopic life-form of *Physarum*.

Now we return to the illegitimate, rare plasmodia, which normally are not mutants, as they have non-selfing progeny. We recall that these are the rare selfers (1 in 10^9), they are not homothallic strains as they do not arise by fusion and do not change the ploidy in their life cycle. Of course, if one takes such amoebae and grows them in clones, eventually one will find some that do manage to make plasmodia in clones. These are examples of real selfers, i.e. mutants with progeny that can self. They have an appropriate name – 'gad' – standing for greater asexual development (Adler & Holt, 1977; Shinnick & Holt, 1977). These mutants are temperature-sensitive, i.e. they self at the permissive temperature. As it has turned out, the famous strain Colonia is really a gad mutant, having the parental gene *A* allele (mt_2) that is typical for the Wisconsin strain (maybe this specimen somehow got from Wisconsin to the botanical garden in Cologne a long time ago).

When a gad mutant is crossed appropriately, it is found that these mutants have retained their parental gene *A* allele (in Colonia it happens to be mt_2) and, as usual, by now most are very tightly linked (one of them could not be separated from the gene *A* in over 5 000 clones tested). Some of the gad mutants are closely linked to the *A* gene, yet clearly separable (about 12 map units apart), one mutant was unlinked, defining perhaps a new gene that supports selfing.

If we pause for a moment, these results confirm the importance of the complex gene A locus and at the same time widen it a little to incorporate those genes that are closeby (quite close, and too close for random distribution of dif$^-$ function), and we get yet another unlinked mutant. We may begin to worry that the formal genetics gets too complex. Maybe, the complex locus, while being formally defined, does not exist at all (we will come back to this, later).

For now, let us turn to yet another set of selfers that were produced in the laboratory. They have been named het$^-$ (for not being heterothallic any more) by Honey (1979). In three cases, by test crossing it has been found that the parental gene A allele has not been changed, as in all other developmental mutants described so far. They are just some more gad mutants.

However, in one particular het$^-$-mutant described by Honey (1979), the parental gene A allele has been no longer found in the progeny. This particular mutant (derived from its parental gene A allele and thus supposedly different from it) crossed with its parent (still having the same A allele) successfully. In contrast to all other mutants, its progeny never reveal the parental A allele next to the mutant A allele (a situation postulated by the $A \neq$ dogma). Therefore, the mutant must be inside the parental A allele. This is the first example of a true parental gene A mutation (the mating-type locus that is different in each isolate found in nature so far, anyhow). It also tells us that among hundreds of mutants analysed it happens very rarely. This could mean one of three things: (1) the parental allele is highly protected (thus very stable), or (2) it gets fixed very fast, after having been injured in mutagenesis, or (3) it does not exist as such and has to get made in each generation, i.e. created again and again. It is clear that three alternatives are two too many and that the third one is heretic. Here, at once, the yeast mating-type system comes to mind, and fungi have tried to become macroscopic creatures, too.

Finally, of course, people have looked for revertants of the gad mutants, i.e. tried to detect individuals that cannot self any more. About 50 'revertants' (gad$^-$) have been found but all make plasmodia more frequently than the original strains. Thus, none of them seems to be a real back mutation. They have been classified as 'suppressor mutants', which somehow interfere with the selfing function (i.e. gad$^-$). Interestingly, even all of these mutants turn out to be linked to the gene A complex (Holt, personal communication). We have encountered a similar phenotype above, where we detected dif$^-$ mutants, originating from selfing strains (defined dif$^+$). All these revertants have the same phenotype: they cannot make plasmodia in amoebal clones, they need to mate (except for very rare illegitimate plasmodium formation). An interesting question has been asked: how do the various revertants (dif⁻) compare with the gad suppressor mutants (gad⁻), when each is crossed with the other? (Remember, there were four complementation groups among the dif (dif⁻) mutants.) Many of the gad suppressor mutants fit the two main complementation groups that were closely linked to, if not allelic with, the complex gene *A* locus (Anderson, 1979). In addition, two new complementation groups have been defined, making a total of six different complementation groups, four very much linked to gene *A* and possibly within the same locus. This sounds quite simple, as one can saturate the plasmodial formation with one very big or maximally six not so big 'genes'. However, it remains a very confused picture, as long as we rely solely on formal genetics. As we look over other complex genetic loci, take into account some recent progress in molecular genetics, and allow some – admittedly poorly founded – speculation in constructing a model, plasmodium development turns out to be a network of just a few events. Some are essential, some not, some are restricted to slime mould plasmodia, some perhaps ubiquitous and basic to all multicellular life-forms and even amenable to experimental analysis. Main trouble is, we have no idea of what the gene products of *mat A* (A ≠ paradigm) look like; let us settle for x and y.

Models of moulding moulds and things

How does an organism rise above unicellularity? Of course, just that happens when egg and sperm fuse and an animal develops from a tiny zygote. It is not clear how multicellularity has evolved. In theory, several routes seem possible.

(1) Unicellular organisms (many believe the choano flagellates) might have come together, formed an aggregate, and divided labour between reproductive cells and those that specialise in feeding and protecting. The genes responsible for these alternative developmental pathways would have to have been preserved and carefully protected during the life time of the reproductive cells (the gametes), as the other cells (committed to somatic functions) could not easily assume generative functions again. The cellular slime moulds can be viewed as examples of aggregation as part of the life cycle.

(2) As a parallel pathway, a gene may have evolved (the mating gene *B*) which permitted fusion of two unicellular but otherwise genetically identical cells. This would have led to a fused creature, or synkaryon, which has two nuclei in a common cytoplasm. If those nuclei went on dividing without fusion, and cell membranes were put in, this creature would get bigger: a dikaryotic form of organism would have been created, as in the fungi.

We have already mentioned that the mating-type gene in yeast is a bipolar system. However, it is very labile. Consequently, the yeast cells can easily get switched from a male to a female back and forth during vegetative growth. As a reminder, the mating gene in *Physarum*, the *mat B* locus, is multi-allelic (myxamoebae do not come in only two forms) and not too stable. In contrast, higher organisms are either male or female in most cases, though not in all.

All these organisms, however, arise only when the two nuclei also fuse with one another after cell fusion. This creates a zygote and initiates the diploid phase of the life cycle. It also requires meiosis to get back to the gamete state. However, the zygote must not lead to a complex organism automatically. Although nuclear fusion precedes growth and differentiation, say in man, it can also be the foreplay to meiosis right away as in amoebae, flagellates and yeast, and not the creation of the multicellular state. And in *Physarum*, as we have seen, even fusion of nuclei after amoebal fusion (being $B\neq$) only creates more amoebae (although they are now diploid). There must be something else, and there is.

(3) Another kind of gene must have evolved which, when activated, changed the state of a unicellular organism such that it could grow bigger and become more complex and sometimes beautiful. This gene then provided the basic switch from microscopic unicellular to macroscopic multicellular life; it was a major developmental gene. About that gene in *Physarum*, the *mat A* locus, we have just given some evidence from formal genetics which describes it as 'complex'. We need one more function (4) before we can describe a life cycle as simple as that of *Physarum* or as complex as ours: a sex function. While the third gene, the developmental switch gene, is not operating in unicellular organisms, the sex gene works in those creatures as well. Its function can only be described in that – after fusion of the gametes – the nuclei undergo the right kind of fusion, karyogamy, that leads to a zygote. Maybe, a better description of that sex gene is that it switches a unicellular state to a gamete. This could account for amoeba–gamete transition in *Physarum* (or germ cell-programming during oogenesis generally) and both may be irreversible commitments. The fact that a somatic nucleus in the famous transplantation experiments with *Xenopus*, or the mouse, can replace the egg nucleus, tells us two things: (1) that the gamete state, once attained, is a function of its cytoplasm and (2) that the main function of the postulated sex-gene may not be fusion of haploid nuclei at all, but a stable alteration of nuclear functions (an instruction). This is obvious for the many 'selfers' in *Physarum* as they require no nuclear fusion. Actually, that sex-function may be one form of *mat A* complex.

Of course, these (and probably some more) genes could act in one and the same organism at some time, leading to the typical sequence of events like: gamete formation, fusion of two sexually distinct gametes, zygote formation with gene re-shuffling, and development (i.e. growth and differentiation and meiosis). Various timing and coupling of these events allows one to describe the many different life cycles that are found in Nature. This situation may well suit *Physarum* development, as we know that gene *mat B* controls fusion of its gametes, while the other multi-allelic gene *mat A* controls the creation of the macroscopic plasmodium. The uncoupling of the function of the B gene (leading to gamete fusion), from the A gene, and the distinction of the selfer from the non-selfer, may provide a good test system to understand one, maybe the only, basis of multicellularity.

Let us first summarise some of the events that distinguish a plasmodium from a gamete (Fig. 60). Provided both genes are present in different alleles in the two gametes ($A \neq$ and $B \neq$), they fuse. Then their nuclei fuse (in interphase!), and a preplasmodium has been created. Centrioles disappear, the intranuclear spindle is organised by a dividing MTOC (microtubule organising centre), followed by closed mitosis. Cell division no longer occurs (maybe due to a self-produced inhibitor that is made along with the slime), and young plasmodia ignore their amoebae, or eat them as well as their cysts, yet recognise other, i.e. older, plasmodia. If they fuse with each other, only at that time must the fusion genes have become expressed, otherwise the big plasmodium would reject the little one or kill it. In this group of early events, cell fusion and nuclear fusion must follow each other, which is self-evident. All other events, however, can occur in any sequence. Their order does not appear to be fixed. Later events, typical for the plasmodium, comprise the shuttle streaming, the granules and their yellow pigment, synchronous mitosis, and finally sporulation. They characterise an established slime mould plasmodium, while the early events mark the beginning of a developmental pathway. If the selfer (or the apogamic or asexual gamete) really follows the same line as heterothallic *Physarum* myxamoebae, the picture may get simpler as we dissect the functions of gene *A* from gene *B*.

If, in two clones of amoebae the mating gene *B* is different ($B \neq$), yet the developmental gene *A* is not ($A =$), cell fusion of amoebae takes place, yet nuclear fusion is later (in prophase not interphase) and either dikaryotic cells or diploid amoebae result, both of which can go on proliferating. This is a puzzling result as in Nature, even in one single fruiting body, under the limitations posed by the tetrapolar mating system in heterothallic strains (gene B_1/B_2 and gene Ax/Ay), 50% of the matings will be unsuccessful and 50% cannot mate in the first place (being $B \neq$); this follows because A and B genes are not linked, and they will assort at random in meiosis. However, with our very limited understanding of the state of 'competence' of a gamete, it might well be that in the natural habitat yet another gene, like *imz* (ionic modulation of zygote formation), may become expressed properly only if both the mating gene *B* and the developmental gene *A* are present in different alleles ($B \neq$ and $A \neq$), and thus indirectly control mating.

There are some arguments that the selfers (the self-fertile, non-sexual strains) do follow the same developmental path as the heterothallic strains

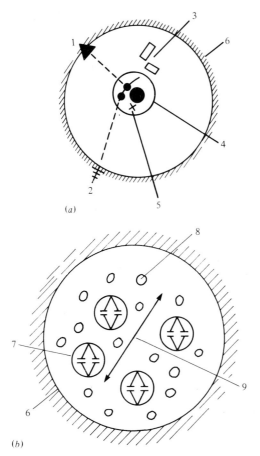

Fig. 60. Some alterations upon plasmodium formation. (*a*) Early changes include: repression of sexual fusion (*mat B* function) (1), expression of plasmodial fusion genes (*fus*-gene function) (2), disappearance of centrioles (3), closed mitosis with intranuclear MTOC (4), secretion of slime (5). (*b*) Later changes: inhibition of cell division (6), mitotic synchrony (7), yellow pigment granules (8), shuttle streaming (9), capacity to sporulate.

(which require sexual fusion). First, most mutants that change a sexual strain into a non-sexual one (or vice versa, i.e. change a selfer into a sexual strain) are located at almost the same place, the *A* locus, or very close by (yet, with one exception, never exactly at the same site). Second, in both types of plasmodial development gametes need a lag-phase to become competent. Third, the inducer substance (even if secreted from a culture of selfers) stimulates the frequency of plasmodium formation in selfing as well as in heterothallic gametes.

Formally, we can argue that a gamete can be induced to differentiate into a plasmodium only if two different alleles of gene A ($A \neq$ or Ax/Ay) are expressed. This also means that a single A allele or two identical ones may repress plasmodium formation. As the heterothallic strains prevail in Nature, at least for *Physarum*, the normal situation, say wild type, means that development of plasmodium is repressed, and two steps must happen: first gamete fusion, then zygote formation. One can argue that cell fusion causes rearrangement in the hybrid cell membrane (now carrying two different gene B products), which in turn causes derepression (or induction) of the plasmodium differentiation pathway. This option is available only if two different A genes (or their products) occur in the same cell (Fig. 61a). Then the two gene functions can complement each other. They could establish a heteromeric structure (Ax/Ay, Ax/Ay . . .), in the cytoplasm or even at the membrane, which then controls indirectly nuclear functions in a new direction. Here, the induction of plasmodium formation would have become an epigenetic process, far away from the genome, as the same gene products could have been around in the two separate cells for some time (Fig. 61a). On the other hand, the homogeneous gene A product could act as a repressor on the chromatin in one gamete, which becomes removed as it gets complexed with the allelic gene A product after mating. Switching of gene activity could immediately follow, controlling plasmodium development at the level of the genome (Fig. 61b). As we cannot distinguish in time sequence the early events during initiation of plasmodium formation which occur at the membrane (recognition) of two gametes and at the nucleus during nuclear fusion (or nuclear swelling, in the selfer where no mating takes place), this important alternative remains open. But we are reminded that these arguments have also occurred in the genetic and epigenetic discussion surrounding the differentiation process of the mature plasmodium.

How can we explain the action of the many mutants close to, at or inside the complex gene A? If we start from a heterothallic strain, mutants interfere with the wild type in that inhibition of plasmodium development no longer prevails (or development of plasmodium becomes repressed by the mutant of a selfer). Whereas the wild type has to become heterogeneous with respect to two gene A alleles via cell fusion, the selfing amoeba, after it has become a gamete, needs to express two alleles of gene A. This ability is provided by the 'mutation' close by the original, i.e. parental gene A, which is still there and does not get changed. One way to explain these 'mutations' is that they are really an activation of a 'silent' gene A allele that has been there all along. Then, there would be two different gene A functions in one cell and no mating would be required for amoeba–plasmodium transition (*apt*), nor would meiosis be a requirement for plasmodium–amoeba transition (*pat*; Fig. 61c). We still do not know whether the activation of a 'silent' gene acts first on the periphery and then on the nucleus (remember that the nucleus becomes big – pregnant, so to speak – just before closed mitosis is

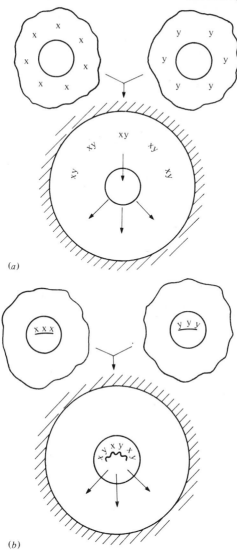

Fig. 61. Possible, yet purely hypothetical, functions of the complex *mat A* locus in transformation of a microscopic cell into a macroscopic plasmodium. (*a,b*) In heterothallic strains, an amoeba expresses either one or another allele of gene *mat A*, say the product x or y which forms a homopolymer (that may repress plasmodium formation of an amoeba). After sexual fusion (*mat B* =), a heteropolymer product (x, y) is formed (which may activate the zygote state and plasmodia formation). At least two possibilities remain: (*a*) the *mat A* product and the heteropolymer (xy) appear outside the nucleus and affect the genome indirectly, or (*b*) the gene products and the heteropolymer (xy) affect genome expression directly.

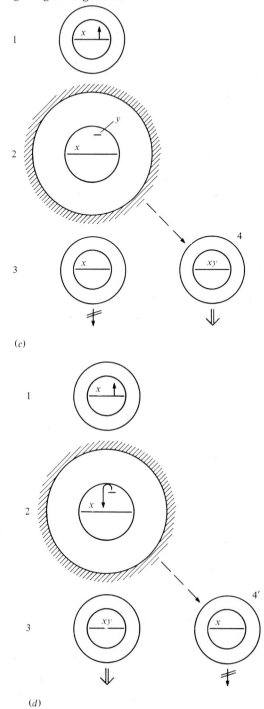

(c)

(d)

observed). We also cannot distinguish whether the inducing factor, which is made by a large crowd of amoebae, really turns some of them into gametes or acts later on during plasmodium initiation.

This interpretation of an activated gene is consistent with the regular creation of a 'selfer' when diploid amoebae are produced (by meiosis in a tetraploid plasmodium), which can turn directly into a plasmodium. One would assume that two different alleles of gene A can interact and induce plasmodium formation. However, these plasmodia are not self-fertile. This means that after they undergo meiosis their progeny cannot self any more. Why not? One could argue that the nuclei that expressed both alleles of gene A, after meiosis contain only one allele of gene A and thus the ensuing amoebae have to be heterothallic again.

Now we still need to understand the illegitimate plasmodia (those that are selfers, though very rare, but that are not self-fertile, as their amoebae, which have not changed their ploidy level, cannot self). If we speculate that 'activation' of a new A allele really means excision of one from a cassette of many and putting it where it can be expressed, we can come up with a formal explanation. The new gene A allele, though active in initiating a plasmodium, does not remain in the plasmodium. It may get eliminated (during proliferation of plasmoidal nuclei, during nuclear elimination, or at plasmodium–amoebae transition, or at gamete formation), i.e. this A allele is labile.

Coming back to the true, i.e. self-fertile, selfer: putative excision of an A allele out of a cassette and integration right next to the parental allele could stabilise this situation, leading to the $A \neq$ paradigm of plasmodium initiation, even at the genome level. A mutant of this linkage would create a dif$^-$ phenotype and a back-insertion of another cassette would restore the dif$^+$ mutant again (Fig. 61d).

Having employed DNA transposition as a hypothetical vehicle to explain some of the developmental mutants in *Physarum*, and having stressed many times through this book the significance of irreversible commitments in the life cycle of *Physarum*, one can ask whether such events at the DNA level

---

Fig. 61. (*cont.*) (*c,d*) The selfing strain may arise by an activation of an 'inactive' allele of *mat A*, by an unknown mechanism. (*c*) In the illegitimate plasmodium formation, a 'silent' allele, say y, becomes activated (1, arrow), yet this state is not stable in the plasmodium (2), leading to progeny (3) with only the maternal allele x of *mat A*, which are non-selfers (i.e. heterothallic) (broken arrow). In a rare event, both alleles x and y become stably arranged and regularly expressed in the gamete (4); then a gad mutation has occurred, and progeny can self (open arrow). (*d*) A selfer is believed to arise, when the activation of a silent allele y of *mat A* in a gamete (1) becomes stabilised in the plasmodium (2), so that the progeny (3) will also be selfers (open arrow), i.e. expressing x and y in the same cell. If that state of *mat A* becomes altered by a mutation, a dif$^-$ strain arises (4'). Phenotypically, this dif mutant is a heterothallic strain (it cannot self, broken arrow), just as the gad mutant is a true selfer.

(that the geneticist in rare cases detects as mutants) are really quite frequent – even essential – and just go unnoticed because it is normal. It could turn out that what we have been calling a gene *A* allele is really a composition of two pieces of DNA that need to be physically linked up at each initiation of a new generation. Although there is no need to postulate this for the heterothallic amoebae, this may occur just as in the selfer or at every act of irreversible commitment in any living thing. We do not expect random scrambling over the genome, rather localised delicate events at a very small fraction of the genome; maybe the same pieces of the genome that determine the yeast mating factor, germ cells in animals, or code for the embryonic surface antigens, or the major histocompatibility functions, the bithorax gene in *Drosophila* pattern formation, the constitution of the immunoglobin genes, and the human Hy antigen. While this is pure speculation, it would be interesting sometime to compare the respective cloned DNA squences with those of gene *A* from *Physarum*. In support of irreversible changes at commitment we can once again cite that nuclear transplantation experiments have never shown a complete reversion of the differentiated state.

It may be a while before this speculation can be discarded, but in the meantime there is much ground to be covered, combining elaborate genetic analysis of plasmodium formation with good biochemical work.

While we may wonder for ever how man develops from a tiny egg cell, we may get a glimpse from the slime mould of how it may begin.

Conclusion

On our way through the lives of *Physarum*, we have encountered many, if not all, major phenomena of developmental biology, although I would not claim *Physarum*, the multiheaded, to be the model system for neurobiology!

We have not arrived at many clear-cut answers, as we have not been able to come up with many hard facts on which all researchers involved would consent. However, the life cycle of *Physarum* contains a great potential for the immediate future now that the basic work has been done and several major concepts have been displayed.

The main advantage lies in the unique possibility to separate growth from differentiation, and the microscopic from the macroscopic life-forms, and to employ immense numbers of haploid organisms for mutagenesis work, enabling genetic dissection of the life cycle and the mitotic cycle as well.

The developmental biologist will have recognised all the mysterious events attributed to becoming tall and different, in particular during the irreversible sporulation process: initiation, competence, induction, commitment, polarity, pattern formation, morphogenesis, and cytodifferentation, and possibly quantal cell cycles.

A unique opportunity is presented by the mechanism of the *mat A* gene complex which may be the major switch from unicellular to multicellular life. Now, the molecular geneticist must enter the field, and isolate that gene, and work out its regulation. I would not be surprised if we depend on its action as well.

Another area which needs to be explored further in the microscopic stages is the factors produced by the amoebae that turn them into gametes and allow them to communicate with each other.

While it is futile to worry about whether *Physarum* is more an animal than a plant, the amoeba–flagellate transition provides an excellent approach to unravel intracellular architecture, and 'what happens to the MTOC?' is a much more relevant question than arguing whether there is a 'differentiation' of a drop of endoplasm within a few minutes into an oscillating protoplast.

Cell biologists should have much opportunity to study oscillations in the shuttle streaming, and they need to find out what causes the exact synchrony in contraction every couple of minutes and what initiates mitosis every eight hours.

Also, one would like to know how the fusion genes provide a recognition mechanism of self, and how the plasmodium can sense its surroundings and detect food, light and potential dangers.

The naturally synchronous mitotic cycle of the plasmodium offers innumerable approaches and, who knows, the G_1^- state of the cell cycle may be the normal proliferative or free-wheeling cell cycle *par excellence*.

Chromatin structure, while displaying all bona fide eukaryotic features, seems to be arranged such that active fractions may be selectively analysed, and of course the accurate initiation of DNA replication will attract more molecular biologists, once a bank of DNA genes is freely available. For the gene expression specialist, the sequential replication and potential replication–transcription coupling may hold many a surprise in finding out how genes work.

The minichromosome with its palindromic rDNA genes is a system of ingenious simplicity to study one gene, which comes complete with all regulatory signals and putative transcription factors, and we still need to know whether it is a 'symbiont' or derived from a chromosomal nuclear gene.

This may tell us where *Physarum* came from in the evolutionary journey of organisms, and in the demonstrated nuclear elimination process one may wonder whether the plasmodium, the unique form of life, is really a playground where the forces of evolution are at work under our eyes and ready to be comprehended as they mould a mould. We may be even in the position to build one, because that is all that developmental biology is about. Let us find out!

References

Achenbach, U. & Wohlfarth-Bottermann, K. E. (1980). Oscillating contractions in protoplasmic strands of *Physarum*. *J. Exp. Biol.* **85**, 21–31.

Adams, D. S., Noonan, D. & Jeffery, W. R. (1980). An improved method for the isolation of polysomes from synchronous macroplasmodia of *Physarum polycephalum*. *Analyt. Biochem.* **103**, 408–12.

Adler, P. N. & Holt, C. E. (1977). Mutations increasing asexual plasmodium formation in *Physarum polycephalum*. *Genetics*, **87**, 401–20.

Aldrich, H. C. (1968). The development of flagella in swarm cells of the myxomycete *Physarum flavicomum*. *J. Gen. Microbiol.* **50**, 217–22.

Aldrich, H. C. & Blackwell, M. (1976). Resistant structures in the myxomycetes. In *The Fungal Spore: Form and Function*, ed. D. J. Weber & W. M. Hess, pp. 413–61. Wiley Interscience, New York.

Aldrich, H. C. & Daniel, J. W. (eds.) (1982). Cell Biology of *Physarum* and *Didymium*. Monographs in Cell Biology Series. Academic Press, London & New York, in press.

Alexopoulos, C. J. (1960). Gross morphology of the plasmodium and its possible significance in the relationships among myxomycetes. *Mycologia*, **52**, 1–20.

Anderson, J. D. (1964). Regional differences in ion concentration in migrating plasmodia. In *Primitive Motile Systems in Cell Biology*, ed. R. D. Allen & N. Kamiya, pp. 125–36. Academic Press, New York.

Anderson, R. W. (1977). A plasmodial colour mutation in the myxomycete *Physarum polycephalum*. *Genet. Res., Camb.* **30**, 301–6.

Anderson, R. W. (1979). Complementation of amoebal–plasmodial transition mutants in *Physarum polycephalum*. *Genetics* **91**, 409–19.

Ashworth, J. M. & Dee, J. (1975). *The Biology of Slime Moulds*. Edward Arnold, London.

Babcock, K. L. & Mohberg, J. (1975). The pedigree of Wis 1 *Physarum polycephalum*. *Physarum Newsletter*, **7**, 3–5.

Baeckmann, A. (1980). Bestimmung der Sequenzkomplexität der poly(A)-RNA von *Physarum polycephalum*. PhD Thesis, University of Würzburg.

Baranetzki, J. (1876). Influence de la lumière sur les plasmodia des myxomycetes. *Mem. Soc. Sci. Nat. Cherbourg*, **19**, 321–60.

Beach, D., Piper, M. & Shall, S. (1980). Isolation of newly-initiated DNA from the early S-phase of the synchronous eukaryote *Physarum polycephalum*. *Exp. Cell Res.* **129**, 211–23.

Bélanger, G., Bellemare, G. & Lemieux, G. (1979). Ribosomal phosphoproteins in *Physarum polycephalum*. *Biochem. Biophys. Res. Commun.* **86** (3), 862–8.

Bernstam, V. A. (1978). Heat effects on protein biosynthesis. *Ann. Rev. Plant. Physiol.* **29**, 25–46.

Berridge, M. J. & Rapp, P. E. (1979). A comparative survey of the function, mechanism and control of cellular oscillators. *J. Exp. Biol.* **81**, 217–79.

Bialczyk, J. (1979). An action spectrum for light avoidance by *Physarum nudum* plasmodia. *Photochem. Photobiol.* **30**, 301–3.

Blessing, J. & Lempp, H. (1978). An immunological approach to the isolation of factors with mitotic activity from the plasmodial stage of the myxomycete *Physarum polycephalum*. *Exp. Cell. Res.* **113**, 435–8.

Block, I. & Wohlfarth-Bottermann, K. E. (1981). Blue light as a medium to influence oscillatory contraction frequency in Physarum. *Cell Biol. Intern. Rep.*, **5**, 73–81.

Bohnert, H. J. (1977). Size and structure of mitochondrial DNA from *Physarum polycephalum*. *Exp. Cell Res.* **106**, 426–8.

Bonner, J. T. (1967). *The Cellular Slime Moulds*, 2nd edn. Princeton University Press, New Jersey.

Bradbury, E. M., Inglis, R. J., Matthews, H. R. & Langan, T. A. (1974). Molecular basis of control of mitotic cell division in eukaryotes. *Nature, London*, **249**, 553–6.

Brand, G., Hüttermann, A. & Haugli, F. (1975). Differential expression of RNase activities in the life cycle of *Physarum polycephalum*. *Naturwissenschaften*, **62**, 535–6.

Brandza, M. (1926). Sur l'influence de la chaleur et de l'evaporation rapide sur les myxomycètes calcarées vivant en plein soleil. *C.R. Acad. Sci. (Paris)*, **182**, 488–9.

Braun, R., Hall, L., Schwärzler, M. & Smith, S. S. (1977). The mitotic cycle of *Physarum polycephalum*. In *Cell Differentiation in Microorganisms, Plants and Animals*, ed. L. Nover & K. Mothes, pp. 402–23. VEB Gustav Fischer Verlag, Jena.

Braun, R., Mittermayer, C. & Rusch, H. P. (1965). Sequential temporal replication of DNA in *Physarum polycephalum*. *Proc. Natl. Acad. Sci., USA*, **53**, 924.

Brewer, E. N. & Busacca, A. (1979). DNA-synthesis in a subnuclear preparation isolated from *Physarum polycephalum*. *Biochem. Biophys. Res. Commun.* **91**, 1352–7.

Brewer, E. N. & Rusch, H. P. (1966). Control of DNA replication: effect of spermine on DNA polymerase activity in nuclei isolated from *Physarum polycephalum*. *Biochem. Biophys. Res. Commun.* **25**, 579–84.

Brewer, E. N. & Rusch, H. P. (1968). Effect of elevated temperature shocks on mitosis and on the initiation of DNA replication in *Physarum polycephalum*. *Exp. Cell Res.* **49**, 79–86.

Brewer, E. N. & Ting, P. (1975). DNA replication in homogenates of *Physarum polycephalum*. *J. Cell. Physiol.* **86**, 459–70.

Burland, T. G. (1978). Temperature-sensitive mutants of *Physarum polycephalum*, a search for cell-cycle mutants. PhD Thesis, University of Leicester.

Burland, T. G., Chainey, A. M., Dee, J. & Foxon, J. L. (1981). Analysis of development and growth in a mutant of *Physarum polycephalum* with defective cell division. *Devel. Biol.* **85**, 26–38.

Burland, T. G. & Dee, J. (1979). Temperature-sensitive mutants of *Physarum polycephalum* expression of mutations in amoebae and plasmodia. *Genet. Res., Camb.* **34**, 33–40.

Burland, T. G. & Dee, J. (1980). Isolation of cell cycle mutants of *Physarum polycephalum*. *Molec. Gen. Genet.* **179**, 43–8.

Carlile, M. J. (1970). Nutrition and chemotaxis in the myxomycete *Physarum polycephalum*: the effect of carbohydrates on the plasmodium. *J. Gen. Microbiol.* **63**, 221–6.

Carlile, M. J. (1973). Cell fusion and somatic incompatibility in myxomycetes. *Ber. Deut. Bot. Ges.* **86**, 123–39.

Carlile, M. J. (1976). The genetic basis of the incompatibility reaction following plasmodial fusion between different strains of the myxomycete *Physarum polycephalum*. *J. Gen. Microbiol.* **93**, 371–6.

Carlile, M. J. & Dee, J. (1967). Plasmodial fusion and lethal interaction between strains in a myxomycete. *Nature, London*, **215**, 832–4.

Carlile, M. J. & Gooday, G. W. (1978). Cell fusion in myxomycetes and fungi. In *Membrane Fusion*, ed. G. Poste & G. L. Nicolson, pp. 219–65. Elsevier/North-Holland Biomedical Press, Amsterdam & New York.

Chapman, A. & Coote, J. G. (1979). Optimum conditions for sporulation in *Physarum polycephalum* strain CL. In *Current Research on* Physarum, *Proceedings of the 4th European* Physarum *Workshop*, ed. W. Sachsenmaier, pp. 10–13. Publications of the University of Innsbruck 120.

Chet, I. (1973). Changes in RNA during differentiation of *Physarum polycephalum*. *Ber. Deut. Bot. Ges.* **86**, 77–92.

Chet, I. & Hüttermann, A. (1977). Melanin biosynthesis during differentiation of *Physarum polycephalum*. *Biochim. Biophys. Acta*, **499**, 148–55.

Chet, I., Naveh, A. & Henis, Y. (1977). Chemotaxis of *Physarum polycephalum* towards carbohydrates, amino acids and nucleotides. *J. Gen. Microbiol.* **102**, 145–8.

Chet, I., Retig, N. & Henis, Y. (1973). Changes in RNases during differentiation (spherulation) of *Physarum polycephalum*. *Biochim. Biophys. Acta*, **294**, 343–7.

Chet, I. & Rusch, H. P. (1969). Induction of spherule formation in *Physarum polycephalum* by polyols. *J. Bacteriol.* **100**, 674–8.

Cheung, W. Y. (1980). Calmodulin plays a pivotal role in cellular regulation. *Science*, **207**, 19–27.

Chin, B., Friedrich, P. D. & Bernstein, I. A. (1972). Stimulation of mitosis following fusion of plasmodia in the myxomycete *Physarum polycephalum*. *J. Gen. Microbiol.* **71**, 93–101.

Clark, J. (1980). Competition between plasmodial incompatibility phenotypes of the myxomycete *Didymium iridis*. II. Multiple clone crosses. *Mycologia*, **72**, 512–22.

Clark, J. & Collins, O. N. (1973). Directional cytotoxic reactions between incompatible plasmodia of *Didymium iridis*. *Genetics*, **73**, 247–57.

Clark, J. & Hakim, R. (1980*a*). Aging of plasmodial heterokaryons in *Didymium iridis*. *Molec. Gen. Genet.* **178**, 419–22.

Clark, J. & Hakim, R. (1980*b*). Nuclear sieving of *Didymium iridis* plasmodia. *Exp. Mycol.* **4**, 17–22.

Collins, O. R. (1979). Myxomycete biosystematics: some recent developments and future research opportunities. *Bot. Rev.* **45**, 145–201.

Cummins, J. E. & Rusch, H. P. (1966). Limited DNA synthesis in the absence of protein synthesis in *Physarum polycephalum*. *J. Cell Biol.* **31**, 577–83.

Cummins, J. E. & Rusch, H. P. (1968). Natural synchrony in the slime mould *Physarum polycephalum*. *Endeavour*, **27**, 124–9.

Cummins, J. E., Weisfeld, G. E. & Rusch, H. P. (1966). Fluctuations of ^{32}P-distribution in rapidly labeled RNA during the cell cycle of *Physarum polycephalum*. *Biochim. Biophys. Acta*, **129**, 240–8.

Daniel, J. W. (1966). Light-induced synchronous sporulation of a myxomycete. In *Cell Synchrony, Studies in Biosynthetic Regulation*, ed. I. L. Cameron & G. M. Padilla, pp. 117–52. Academic Press, New York.

Daniel, J. W. (1976). Dual control of myxomycete sporulation and growth by photometabolic regulation of c-AMP and calcium-mediated changes in permeability. (Abstr.) *VII International Congress on Photobiology*, Rome.

Daniel, J. W. & Baldwin, H. H. (1964). Methods of culture for plasmodial myxomycetes. In *Methods in Cell Physiology*, ed. D. M. Prescott, vol. 1, pp. 9–41. Academic Press, New York.

Daniel, J. W. & Rusch, H. P. (1961). The pure culture of *Physarum polycephalum* on a partially defined soluble medium. *J. Gen. Microbiol.* **25**, 47–59.

Daniel, J. W. & Rusch, H. P. (1962). Method for inducing sporulation of pure cultures of the myxomycete *Physarum polycephalum*. *J. Bacteriol.* **83**, 234–40.

Davidson, E. H. (1976). *Gene Activity in Early Development*. Academic Press, New York, San Francisco & London.

Davies, K. E. & Walker, I. O. (1977). In-vitro transcription of RNA on nuclei, nucleoli and chromatin from *Physarum polycephalum*. *J. Cell Sci.* **26**, 267–79.

Davies, K. E. & Walker, I. O. (1978). Control of RNA transcription in nuclei and nucleoli of *Physarum polycephalum*. *FEBS Lett.* **86** (2), 303–6.

Dawkins, R. (1976). *The Selfish Gene*. Oxford University Press.

de Bary, A. (1860). Die Mycetozoen, ein Beitrag zur Kenntnis der niedersten Thiere. *Z. wiss. Zool.* **10**, 88.

Dee, J. (1960). A mating-type system in an acellular slime-mould. *Nature, London,* **185**, 780–1.

Dee, J. (1973). Aims and techniques of genetic analysis in *Physarum polycephalum*. *Ber. Deut. Bot. Ges.* **86**, 93–121.

Dee, J. (1975). Slime-moulds in biological research. *Sci. Progr., Oxford*, **62**, 523–42.

Dee, J. (1978). A gene unlinked to mating type affecting crossing between strains of *Physarum polycephalum*. *Genet. Res., Camb.* **31**, 85–92.

Del Castillo, L., Oustrin, M. L. & Wright, M. (1978). Characterization of thermosensitive mutants of *Physarum polycephalum*. *Molec. Gen. Genet.* **164**, 145–54.

Dove, W. F., Laffler, T. G., Chang, M. T., Gross, C. A., Schedl, T., Warren, N. & Gorman, J. A. (1980). Periodic activities in the mitotic cycle of *Physarum polycephalum*. In *Abstracts, Second International Congress on Cell Biology*, Berlin, S 1466, p. 492.

Dove, W. F. & Rusch, H. P. (eds.) (1980). *Growth and Differentiation in* Physarum polycephalum. Princeton University Press, New Jersey.

Dykstra, M. J. & Aldrich, H. C. (1978). Successful demonstration of an elusive cell coat in amoebae. *J. Protozool.* **25** (1), 38–41.

Eliasson, U. (1977). Recent advances in the taxonomy of myxomycetes. *Bot. Notiser.* **130**, 483–92.

Elliott, S. G. & McLaughlin, C. S. (1978). Rate of macromolecular synthesis through the cell cycle of the yeast *Saccharomyces cerevisiae*. *Proc. Natl. Acad. Sci. USA,* **75**, 4384–8.

Ernst, G. H. & Sauer, H. W. (1977). A nuclear elongation factor of transcription from *Physarum polycephalum in vitro*. *Eur. J. Biochem.* **74**, 253–61.

Fischer, S. G. & Laemmli, U. K. (1980). Cell cycle changes in *Physarum polycephalum* histone H1 phosphate: relationship to deoxyribonucleic acid binding and chromosome condensation. *Biochemistry*, **19**, 2240–6.

Forde, B. G. & Sachsenmaier, W. (1979). Oxygen uptake and mitochondrial enzyme activities in the mitotic cycle of *Physarum polycephalum*. *J. Gen. Microbiol.* **115**, 135–43.

Fouquet, H. & Braun, R. (1974). Differential RNA synthesis in the mitotic cycle of *Physarum polycephalum*. *FEBS Lett.* **38**, 184–6.

Fouquet, H. & Sauer, H. W. (1975). Variable redundancy in RNA transcripts isolated in S and G_2 phase of the cell cycle of *Physarum*. *Nature, London*, **255**, 253–5.

Funderud, S., Andreassen, R. & Haugli, F. (1978). Size distribution and maturation of newly replicated DNA through the S and G_2 period of *Physarum polycephalum*. *Cell*, **15**, 1519–26.

Funderud, S., Andreassen, R. & Haugli, F. (1979). DNA replication in *Physarum polycephalum*: electron microscopic and autoradiographic analysis of replicating DNA from defined stages of the S-period. *Nuc. Acid Res.* **6**, 1417–31.

Gall, J. G. (1974). Free ribosomal RNA genes in the macronucleus of *Tetrahymena*. *Proc. Natl. Acad. Sci. USA*, **71**, 3078–81.

Garrod, D. & Ashworth, J. M. (1973). Development of the cellular slime mould *Dictyostelium discoideum. Symp. Soc. Gen. Microbiol.* **23**, 407–37.

Gerisch, G., Malchow, D. & Hess, B. (1974). Cell communication and cyclic-AMP regulation during aggregation of the slime mould *Dictyostelium discoideum*. In *25. Mosbacher Kolloquium Ges. Biol. Chemie*, ed. L. Jaenicke, pp. 279–98. Springer-Verlag, Berlin.

Gerson, D. F. (1978). Intracellular pH and the mitotic cycle in *Physarum* and mammalian cells. In *Cell Cycle Regulation*, ed. J. R. Jeter, I. L. Cameron, G. M. Padilla & A. M. Zimmerman, pp. 105–31. Academic Press, New York, San Francisco & London.

Gingold, E. C., Grant, W. D., Wheals, A. E. & Wren, M. (1976). Temperature-sensitive mutants of the slime mold *Physarum polycephalum*. II. Mutants of the plasmodial phase. *Molec. Gen. Genet.* **149**, 115–19.

Goodman, E. M. (1972). Axenic culture of myxamoebae of the myxomycete *Physarum polycephalum. J. Bacteriol.* **111**, 242–7.

Goodman, E. M. (1980). *Physarum polycephalum*: a review of a model system using a structure–function approach. *Int. Rev. Cytol.* **63**, 1–57.

Goodman, E. M. & Rusch, H. P. (1970). Ultrastructural changes during spherule formation in *Physarum polycephalum. J. Ultrastruct. Res.* **30**, 172–83.

Goodman, E. M., Sauer, H. W., Sauer, L. & Rusch, H. P. (1969). Polyphosphate and other phosphorus compounds during growth and differentiation of *Physarum polycephalum. Can. J. Microbiol.* **15**, 1325–31.

Gorman, J. A., Dove, W. F. & Shaibe, E. (1979). Mutations affecting the initiation of plasmodial development in *Physarum polycephalum. Develop. Genet.* **1**, 47–60.

Gorman, J. A. & Wilkins, A. S. (1980). Developmental phases in the life cycle of *Physarum polycephalum* and related myxomycetes. In *Growth and Differentiation in* Physarum polycephalum, ed. W. F. Dove & H. P. Rusch, pp. 157–201, Princeton University Press, New Jersey.

Grainger, R. M. & Ogle, R. C. (1978). Chromatin structure of the ribosomal RNA genes in *Physarum polycephalum. Chromosoma, Berlin*, **65**, 115–26.

Grant, W. D. (1972). The effect of α-amanitin and $(NH_4)_2SO_4$ on RNA synthesis in nuclei and nucleoli isolated from *Physarum polycephalum* at different times during the cell cycle. *Eur. J. Biochem.* **29**, 94–8.

Gray, W. D. (1938). The effect of light on the fruiting of myxomycetes. *Am. J. Bot.* **25**, 511–22.

Gray, W. D. & Alexopoulos, C. J. (1968). *Biology of the Myxomycetes*. Ronald Press, New York.

Grębecki, A. & Cieślawska, M. (1978). Plasmodium of *Physarum polycephalum* as a synchronous contractile system. *Cytobiologie*, **17**, 335–42.

Gröbner, P. & Sachsenmaier, W. (1976). Thymidine kinase enzyme variants in *Physarum polycephalum*; change of pattern during the synchronous mitotic cycle. *FEBS Lett.* **71**, 181–4.

Grummt, F., Waltl, G., Jantzen, H. M., Hamprecht, K., Huebscher, U. & Kuenzle, C. C. (1979). Diadenosine $5',5'''-P^1,P^4$-tetraphosphate, a ligand of the 57-kilodalton sub-unit of DNA polymerase α. *Proc. Natl. Acad. Sci. USA*, **76**, 6081–5.

Gubler, U., Wyler, T., Seebeck, T. & Braun, R. (1980). Processing of ribosomal precursor RNAs in *Physarum polycephalum*. *Nuc. Acids Res.* **8**, 2647–64.

Guttes, E., Devi, R. & Guttes, S. (1969). Synchronization of mitosis in *Physarum polycephalum* by coalescence of postmitotic and premitotic plasmodial fragments. *Experientia*, **25**, 615–16.

Guttes, E. & Guttes, S. (1964). Mitotic synchrony in the plasmodia of *Physarum polycephalum* and mitotic synchronization by coalescence of microplasmodia. In *Methods in Cell Physiology*, ed. D. M. Prescott, vol. 1, pp. 43–54. Academic Press, New York.

Guttes, S. & Guttes, E. (1968). Regulation of DNA replication in the nuclei of the slime mold *Physarum polycephalum*. Transplantation of nuclei by plasmodial coalescence. *J. Cell Biol.* **37**, 761–72.

Guttes, E., Guttes, S. & Rusch, H. P. (1961). Morphological observations on growth and differentiation of *Physarum polycephalum* grown in pure culture. *Develop. Biol.* **3**, 588–614.

Haars, A., McCullough, C. H. R., Hüttermann, A. & Dee, J. (1978). Regulation of proteolytic enzmes in axenically grown myxamoebae of *Physarum polycephalum*. *Arch. Microbiol.* **118**, 55–60.

Halvorson, H. O. (1977). A review of current models on temporal gene expression in *Saccharomyces cerevisiae*. In *Cell Differentiation in Microorganisms, Plants and Animals*, ed. L. Nover & K. Mothes, pp. 361–76. VEB Gustav Fischer Verlag, Jena.

Hardman, N. & Gillespie, D. A. F. (1980). DNA replication in *Physarum polycephalum*. Analysis of replicating nuclear DNA using the electron microscope. *Eur. J. Biochem.* **106**, 161–7.

Hardman, N., Jack, P. L., Fergie, R. C. & Gerrie, L. M. (1980). Sequence organisation in nuclear DNA from *Physarum polycephalum*. *Eur. J. Biochem.* **103**, 247–57.

Hartwell, L. H. (1974). *Saccharomyces cerevisiae* cell cycle. *Bacteriol. Rev.* **38**, 167–98.

Hatano, S., Ishikawa, H. & Sato, H. (eds.) (1979). *Cell Motility: Molecules and Organisation*. University Park Press, Baltimore, MD.

Hato, M., Ueda, T., Kurihara, K. & Kobatake, Y. (1975). Changes in zeta potential and membrane potential of slime mold *Physarum polycephalum* in response to chemical stimuli. *Biochim. Biophys. Acta*, **426**, 73–80.

Hato, M., Ueda, T., Kurihara, K. & Kobatake, Y. (1976). Phototaxis in true slime mold *Physarum polycephalum*. *Cell Struct. Func.* **1**, 269–78.

Haugli, F. B., Cooke, D. & Sudbery, P. (1980). The genetic approach in the analysis of the biology of *Physarum polycephalum*. In *Growth and Differentiation in Physarum polycephalum*, ed. W. F. Dove & H. P. Rusch, pp. 129–56. Princeton University Press, New Jersey.

Haugli, F. B., Dove, W. F. & Jimenez, A. (1972). Genetics and biochemistry of cycloheximide resistance in *Physarum polycephalum*. *Molec. Gen. Genet.* **118**, 97–107.

Henney, H. R. & Asgari, M. (1975). Growth of the haploid phase of the myxomycete *Physarum flavicomum* in defined minimal medium. *Arch. Microbiol.* **102**, 175–8.

Henney, H. R. & Chu, P. (1977). Differentiation of *Physarum flavicomum*. Metabolic patterns and the role of amino acids in the control of encystment. *Exp. Mycol.* **1**, 41–51.

Henney, M. R. & Henney, H. R. (1968). The mating-type systems of the myxomycetes *Physarum rigidum* and *P. flavicomum*. *J. Gen. Microbiol.* **53**, 321–32.

Hess, B. & Boiteux, A. (1971). Oscillatory phenomena in biochemistry. *Annu. Rev. Biochem.* **40**, 237–58.

Hildebrandt, A. & Sauer, H. W. (1976a). Zur Entwicklungsbiologie von *Physarum*: Modifikationen des RNA-Polymerase-Musters während Wachstum und der Differenzierung. *Verh. Dtsch. Zool. Ges., 1976*, ed. W. Rathmayer, p. 286. Gustav Fischer Verlag, Stuttgart.

Hildebrandt, A. & Sauer, H. W. (1976b). Differential template specificities of nuclear RNA polymerases isolated from *Physarum polycephalum*. *Arch. Biochem. Biophys.* **176**, 214–17.

Hildebrandt, A. & Sauer, H. W. (1976c). Levels of RNA polymerases during the mitotic cycle of *Physarum polycephalum*. *Biochim. Biophys. Acta*, **425**, 316–21.

Hildebrandt, A. & Sauer, H. W. (1977a). Transcription of ribosomal RNA in the life cycle of *Physarum* may be regulated by a specific nucleolar initiation inhibitor. *Biochem. Biophys. Res. Commun.* **74**, 466–72.

Hildebrandt, A. & Sauer, H. W. (1977b). Discrimination of potential and actual RNA polymerase B activity in isolated nuclei during differentiation of *Physarum polycephalum*. *Wilhelm Roux's Arch. Develop. Biol.* **183**, 107.

Hodapp, E. L. (1942). Some factors inducing sclerotisation in mycetozoa. *Biodynamica*, **4**, 33–46.

Hoffmann, W. & Hüttermann, A. (1975). Aminopeptidases of *Physarum polycephalum*. Activity, isoenzyme patterns, and synthesis during differentiation. *J. Biol. Chem.* **250**, 7420–7.

Hohl, H. R. (1976). Myxomycetes. In *The Fungal Spore: Form and Function*, ed. D. J. Weber & W. M. Hess, pp. 463–98. Wiley-Interscience, New York.

Holmes, R. P. & Stewart, P. R. (1979). The isolation of coupled mitochondria from *Physarum polycephalum* and their response to calcium ions. *Biochim. Biophys. Acta*. **545**, 94–105.

Holt, C. E. (1980). The nuclear replication cycle in *Physarum polycephalum*. In *Growth and Differentiation in* Physarum polycephalum, ed. W. F. Dove & H. P. Rusch, pp. 9–63. Princeton University Press, New Jersey.

Holt, C. E. & Hüttermann. A. (1979). Genetic determination of plasmodium formation in *Physarum polycephalum* (Myxomycetes). *Institut für den wiss. Film, Göttingen*. Film Nr. B 1337.

Holtzer, H., Rubinstein, N., Fellini, S., Yeoh, G., Chi, J., Birnbaum, J. & Okayama, M. (1975). Lineages, quantal cell cycles, and the generation of diversity. *Quart. Rev. Biophys.* **8**, 523–57.

Honey, N. K. (1979). The mating-type locus and differentiation in *Physarum polycephalum*. PhD Thesis, University of Otago.

Hosoda, E. (1980). Culture methods and sporulation of *Physarum polycephalum*. *Mycologia*, **72**, 500–4.

Howard, D. K. & Pelc, S. R. (1953). Synthesis of deoxyribonucleic acid in normal and irradiated cells and its relation to chromosome breakage. *Heredity, Suppl.* **6**, 261–73.

Howard, F. L. (1932). Nuclear division in plasmodia of *Physarum*. *Ann. Bot., London*, **46**, 461–77.

Hüttermann, A. (1973). Biochemical events during spherule formation of *Physarum polycephalum. Ber. Deut. Bot. Ges.* **86**, 55–76.

Inglis, R. J., Matthews, H. R. & Bradbury, E. M. (1976). The molecular basis for the control of cell division. In *Radiation and Cellular Control Processes*, ed. J. Kiefer, pp. 240–8. Springer-Verlag, Berlin, Heidelberg & New York.

Jacobson, D. N. & Dove, W. F. (1975). The amoebal cell of *Physarum polycephalum*: colony formation and growth. *Develop. Biol.* **47**, 97–105.

Jacobson, D. N., Johnke, R. M. & Adelman, M. R. (1976). Studies on motility in *Physarum polycephalum*. In *Cell Motility*, ed. R. Goldman, T. Pollard & J. Rosenbaum, pp. 749–70. Cold Spring Harbor Laboratory, New York.

Jaffe, L. F., Robinson, K. R. & Nuccitelli, R. (1975). Calcium currents and gradients as a localizing mechanism. In *Developmental Biology*, ed. D. McMahon & C. F. Fox. ICN-UCLA Symp. Mol. Cell Biol., vol. 2, pp. 135–47. Benjamin. Menlo Park, California.

Jahn, E. (1928). Myxomycetes. In *Die Natürlichen Pflanzenfamilien*, ed. A. Engler & K. Prantl, vol. 2, pp. 304–39. Engelmann, Leipzig.

Jalouzot, R., Briane, D., Ohlenbusch, H. H., Wilhelm, M. L. & Wilhelm, F. X. (1980). Kinetics of nuclease digestion of *Physarum polycephalum* nuclei at different stages of the cell cycle. *Eur. J. Biochem.* **104**, 423–31.

Jalouzot, R. &. Toublan, B. (1981). Multiple effects of 5 mM sodium butyrate on *Physarum polycephalum* macroplasmodia. *Cell Tissue Res.* **214**, 195–200.

Jeffery, W. R. & Rusch, H. P. (1974). Induction of somatic fusion and heterokaryosis in two incompatible strains of *Physarum polycephalum. Dev. Biol.* **39**, 331–5.

Jerzmanowski, A., Staron, K., Fronk, J. & Toczko, K. (1979). DNA is less tightly bound to nucleosomal histones in *Physarum polycephalum* as compared to higher eukaryotes. In *current research on* Physarum, *Proceedings of the 4th European* Physarum *Workshop*, ed. W. Sachsenmaier, p. 45, Publications of the University of Innsbruck 120. Innsbruck University Press.

Jeter, J. R. Jr. & Cameron, I. L. (1974). Acidic nuclear proteins and the cell cycle. In *Acidic Proteins of the Nucleus*, ed. I. L. Cameron & J. R. Jeter, pp. 213–45. Academic Press, New York.

Jeter, J. R., Cameron, I. L., Smith, N. K. R., Steffens, W. L. & Wille, J. J. (1979). Cell cycle fluctuations in concentration of various elements in cytoplasm and in nucleus/chromatin of *Physarum polycephalum. J. Cell Biol.* **83**, 8A.

Jockusch, B. M. (1973). Nuclear proteins in *Physarum polycephalum. Ber. Deutsch. Bot. Ges.* **86**, 39–54.

Jockusch, B. M., Brown, D. F. & Rusch, H. P. (1970). Synthesis of a nuclear protein in G_2-phase. *Biochem. Biophys. Res. Commun.* **38**, 279–83.

Jockusch, B. M., Sauer, H. W., Brown, D. F., Babcock, K. L. & Rusch, H. P. (1970). Differential protein synthesis during sporulation in the slime mould *Physarum polycephalum. J. Bacteriol.* **103**, 356–63.

Johnson, E. M., Campbell, G. R. & Allfrey, V. G. (1979). Different nucleosome structures on transcribing and non-transcribing ribosomal gene sequences. *Science*, **206**, 1192–4.

Jump, J. A. (1954). Studies on sclerotization in *Physarum polycephalum. Am. J. Bot.* **41**, 561–7.

Kamiya, N. (1959). Protoplasmic streaming. *Protoplasmatologia*, **8** (3a), 1–199.

Kauffman, S. & Wille, J. J. (1975). The mitotic oscillator in *Physarum polycephalum. J. Theor. Biol.* **55**, 47–93.

Kerr, N. S. (1960). Flagella formation by myxamoebae of the true slime mould, *Didymium nigripes. J. Protozool.* **12**, 276–8.

Kerr, S. J. (1972). Inhibition of flagellum morphogenesis in the true slime mould *Didymium nigripes. J. Gen. Microbiol.* **72**, 419–27.

Kessler, D., Eisenlohr, L. C., Lathwell, M. J., Huang, J., Taylor, H. C., Godfrey, S. D. & Spady, M. L. (1980). *Physarum* myosin light chain binds calcium. *Cell Motility*, **1**, 63–71.

Kincaid, R. L. & Mansour, T. E. (1978). Chemotaxis toward carbohydrates and amino acids in *Physarum polycephalum. Exp. Cell Res.* **116**, 377–85.

Kleinig, H. (1974). Differentiation of *Physarum polycephalum*: inhibition by alcohols. *Cytobiologie*, **9**, 240–3.

Knowles, D. J. C. & Carlile, M. J. (1978). The chemotactic response of plasmodia of the myxomycete *Physarum polycephalum* to sugars and related compounds. *J. Gen. Microbiol.* **108**, 17–25.

Komnick, H., Stockem, W. & Wohlfarth-Bottermann, K. E. (1973). Cell motility: mechanisms in protoplasmic streaming and ameoboid movement. *Int. Rev. Cytol.* **34**, 169–249.

Krauth, W. & Werner, D. (1979). Analysis of the most tightly bound proteins in eukaryotic DNA. *Biochim. Biophys. Acta*, **564**, 390–401.

Kuhn, I. (1980). Stage-specific antigens of *Physarum polycephalum. Exp. Cell Res.* **127**, 431–4.

Kuznicki, J. & Drabikowski, W. (1979). Purification and properties of the Ca^{2+}-binding modulator protein from *Physarum polycephalum*. In *Current Research on* Physarum, *Proceedings of the 4th European* Physarum *Workshop*, ed. W. Sachsenmaier. Publications of the University of Innsbruck 120. Innsbruck University Press.

Laane, M. M. & Haugli, F. B. (1974). Division centers in mitotic nuclei of *Physarum polycephalum* plasmodia. *Norw. J. Bot.* **21**, 309–18.

Laane, M. M. & Haugli, F. B. (1976). Nuclear behaviour during meiosis in the myxomycete *Physarum polycephalum. Norw. J. Bot.* **23**, 7–21.

Laane, M. M., Haugli, F. B. & Mellum, T. R. (1976). Nuclear behaviour during sporulation and germination in the Colonia strain of *Physarum polycephalum. Norw. J. Bot.* **23**, 177–89.

Laffler, T. G. & Dove, W. F. (1977). Viability of *Physarum polycephalum* spores and ploidy of plasmodial nuclei. *J. Bacteriol.* **131**, 473–6.

Laffler, T. G., Wilkins, A., Selvig, S., Warren, N., Kleinschmidt, A. & Dove, W. F. (1979). Temperature-sensitive mutants of *Physarum polycephalum*: Viability, growth and nuclear replication. *J. Bacteriol.* **138**, 499–504.

LeStourgeon, W. M., Beyer, A. L., Christensen, M. E., Walker, B. W., Poupore, S. M. & Daniels, L. P. (1978). The packaging proteins of core hnRNP particles and the maintenance of proliferative cell states. *Cold Spring Harbor Symp. Quant. Biol.* **42**, 885–98.

Lewis, W. H. (1942). The relation of the viscosity changes of protoplasm to ameboid locomotion and cell division. In *The Structure of Protoplasm*, ed. W. Seifriz, pp. 163–97. The Iowa State College Press, Ames.

Ling, H. & Ling, M. (1974). Genetic control of somatic fusion in a myxomycete. *Heredity*, **32**, 95–104.

Ling, H. & Upadhyaya, K. C. (1974). Cytoplasmic incompatibility studies in the myxomycete *Didymium iridis*: recovery and nuclear survival in heterokaryons. *Am. J. Bot.* **61**, 598–603.

Liskay, R. M. & Prescott, D. M. (1978). Genetic analysis of the G1-period: isolation of mutants (or variants) with a G1-period from a Chinese hamster cell line lacking G1. *Proc. Natl. Acad. Sci. USA*, **75**, 2873–7.

Loidl, P. (1979). Regulation der synchronen Mitose und DNA-Synthese in *Physarum polycephalum*. PhD Thesis, University of Innsbruck.

Loomis, W. F. (1975). *Dictyostelium discoideum: a Developmental System*. Academic Press, New York. 214 pp.

Lunn, A., Cooke, D. & Haugli, F. (1977). Genetics and biochemistry of 5-bromodeoxyuridine resistance in *Physarum polycephalum*. *Genet. Res., Camb.* **30**, 1–12.

Lutkenhaus, J. F., Moore, B. A., Masters, M. & Donachie, W. D. (1979). Individual proteins are synthesized continuously throughout the *Escherichia coli* cell cycle. *J. Bacteriol.* **138**, 352–60.

McCormick, J., Blomquist, J. C. & Rusch, H. P. (1970). Isolation and characterization of a galactosamine wall from spores and spherules of *Physarum polycephalum*. *J. Bacteriol.* **104**, 1119–25.

McCullough, C. H. R., Cooke, D. J., Foxon, S. L., Sudbery, P. E. & Grant, W. D. (1973). Nuclear DNA content and senescence in *Physarum polycephalum*. *Nat. New Biol.* **245**, 263–5.

McCullough, C. H. R. & Dee, J. (1976). Defined and semi-defined media for the growth of *Physarum polycephalum*. *J. Gen. Microbiol.* **95**, 151–8.

McCullough, C. H. R., Dee, J. & Foxon, J. L. (1978). Genetic factors determining the growth of *Physarum polycephalum* amoebae in axenic medium. *J. Gen. Microbiol.* **106**, 297–306.

Magun, B. E. (1974). Characterization of nuclear phosphoproteins in *Physarum polycephalum*. In *Acidic Proteins of the Nucleus*, ed. I. L. Cameron & J. R. Jeter, Jr. pp. 137–58. Academic Press, New York.

Magun, B. E. (1976). Cytoplasmic DNA-binding phosphoproteins of *Physarum polycephalum*. *Exp. Cell Res.* **103**, 219–31.

Magun, B. E. (1979). Changes in cytoplasmic DNA-binding phosphoproteins during the cell cycle of *Physarum polycephalum*. *Cell Diff.* **8**, 157–72.

Magun, B. E., Burgess, R. R. & Rusch, H. P. (1975). Nuclear phosphoproteins of *Physarum polycephalum*. Characterization and phosphorus content of the phenol-soluble nuclear acidic proteins. *Arch. Biochem. Biophys.* **170**, 49–60.

Martin, G. W. (1940). The myxomycetes. *Bot. Rev.* **6**, 356–88.

Martin, G. W. & Alexopoulos, C. J. (1969). *The Myxomycetes*. University of Iowa Press.

Matsumoto, S. (1977). Onset time of signal for mitosis estimated from mitotic delays in UV irradiated plasmodia of *Physarum polycephalum*. *Cell Struct. Funct.* **2** (2), 101–10.

Matsumoto, S. & Funakoshi, H. (1978). Coupled increases in nuclear and nucleolar sizes with the cell phase transition in the cell cycle of *Physarum polycephalum*. *Cell Struct. Funct.* **3**, 173–80.

Matthews, H. R. & Bradbury, E. M. (1978). The role of histone phosphorylation in the cell cycle. *Exp. Cell Res.* **111**, 343–51.

Matthews, H. R., Swaron, S., Chabal, S. S., Miller, S., Inglis, R. J. & Bradbury, E. M. (1979). *Physarum* chromatin. In *Current Research on* Physarum, *Proceedings of the 4th European* Physarum *Workshop*, ed. W. Sachsenmaier, pp. 51–7. Publications of the University of Innsbruck 120. Innsbruck University Press.

Maynard Smith, J. (1971). What use is sex? *J. Theor. Biol.* **30**, 319–35.

Melera, P. W. (1980). Transcription in the myxomycete *Physarum polycephalum*. In *Growth and Differentiation in* Physarum polycephalum, ed. W. F. Dove & H. P. Rusch, pp. 64–97. Princeton University Press, New Jersey.

Melera, P. W., Davide, J. P. & Hession, C. (1979). Identification of mRNA in the slime mold *Physarum polycephalum*. *Eur. J. Biochem.* **96**, 373–8.

Melera, P. W., Momeni, C. & Rusch, H. P. (1974). Analysis of isoaccepting tRNAs during the growth phase mitotic cycle of *Physarum polycephalum*. *Biochemistry*, **13**, 4139–42.

Melera, P. W. & Rusch, H. P. (1973). Aminoacylation of transfer ribonucleic acid *in vitro* during the mitotic cycle of *Physarum polycephalum*. *Biochemistry*, **12**, 1307–11.

Miller, D. M. & Anderson, J. D. (1966). The morphology, migration and pressure development of oriented plasmodia of the slime mold. *Trans. Illinois Acad. Sci.* **59**, 352–7.

Mir, L., Del Castillo, L. & Wright, M. (1979). Isolation of *Physarum polycephalum* amoebal mutants defective in flagellation and associated morphogenetic processes. *FEMS Microbiol. Lett.* **5** (1), 43–6.

Mitchell, J. L. A. & Rusch, H. P. (1973). Regulation of polyamine synthesis in *Physarum polycephalum* during growth and differentiation. *Biochim. Biophys. Acta.* **297**, 503–16.

Mitchelson, K. R., Bekers, A. G. M. & Wanka, F. (1979). Isolation of residual protein structure from nuclei of the myxomycete *Physarum polycephalum*. *J. Cell Sci.* **39**, 247–56.

Mitchelson, K., Chambers, T., Bradbury, E. M. & Matthews, H. R. (1978). Activation of histone kinase in G2-phase of the cell cycle in *Physarum polycephalum*. *FEBS Lett.* **92**, 339–42.

Mitchison, J. M. (1971). *The Biology of the Cell Cycle*. Cambridge University Press.

Mitchison, J. M. (1977*a*). Enzyme synthesis during the cell cycle. In *Cell Differentiation in Microorganisms, Plants and Animals*, ed. L. Nover, K. Mothes, pp. 377–401. VEB Gustav Fischer Verlag, Jena.

Mitchison, J. M. (1977*b*). The timing of cell cycle events. In *Mitosis Facts and Questions*, ed. M. Little, N. Pawletz, C. Petzelt, H. Postingl, D. Schroeter & H. P. Zimmermann. Springer-Verlag, Berlin, Heidelberg & New York.

Mohberg, J. (1974). The nucleus of the plasmodial slime molds. In *The Cell Nucleus*, ed. H. Busch, vol. 1, pp. 187–219. Academic Press, New York.

Mohberg, J. (1977). Nuclear DNA content and chromosome numbers throughout the life cycle of the Colonia strain of the myxomycete *Physarum polycephalum*. *J. Cell Sci.* **24**, 95–108.

Mohberg, J. & Rusch, H. P. (1970). Nuclear histones in *Physarum polycephalum* during growth and differentiation. *Arch. Biochem. Biophys.* **138**, 418–32.

Mohberg, J. & Rusch, H. P. (1971). Isolation and DNA content of nuclei of *Physarum polycephalum*. *Exp. Cell Res.* **66**, 305–16.

Monroy, A. & Moscona, A. A. (1980). *Introductory Concepts in Developmental Biology*. University of Chicago Press, Chicago & London.

Nachmias, V. T. (1979*a*). From ameba to muscle: on some work by and with John M. Marshall. In *Motility in Cell Function*, ed. F. A. Pepe, J. W. Sanger & V. T. Nachmias, pp. 9–26. Academic Press, New York.

Nachmias, V. T. (1979*b*). The contractile proteins of *Physarum polycephalum* and actin polymerization in plasmodial extracts. In *Cell Motility: Molecules and Organization*, ed. S. Hatano, H. Ishikawa & H. Sato, pp. 33–57. University Park Press, Baltimore, MD.

Nader, W. & Becker, J. U. (1979). 1,4-α-Glucan phosphorylase EC 2.4.1.1. from the slime mold *Physarum polycephalum*. Purification, physicochemical and kinetic properties. *Eur. J. Biochem.* **102** (2), 354–6.

Nasmyth, K. A. & Tatchell, K. (1980). The structure of transposable yeast mating type loci. *Cell*, **19**, 753–64.

Newell, P. C. (1978). The genetics of cellular slime molds. *Annu. Rev. Genet.* **12**, 69–93.

Nicholls, T. J. (1972). The effects of starvation and light on intramitochondrial granules in *Physarum polycephalum. J. Cell Sci.* **10**, 1–14.

Nygaard, O. F., Guttes, S. & Rusch, H. P. (1960). Nucleic acid metabolism in a slime mold with synchronous mitosis. *Biochim. Biophys. Acta*, **38**, 298–306.

Olive, L. S. (1975). *The Mycetozoans.* Academic Press, New York.

Palotta, D. J., Youngman, P. J., Shinnick, T. M. & Holt, C. E. (1979). Kinetics of mating in *Physarum polycephalum. Mycologia*, **71**, 68–84.

Pepe, F. E., Sanger, J. W. & Nachmias, V. T. (eds.) (1979). *Motility in Cell Function.* Academic Press, New York.

Pierron, G. & Sauer, H. W. (1980*a*). RNA polymerase B levels during the cell cycle of *Physarum polycephalum. Wilhelm Roux's Arch. Develop. Biol.* **189**, 165–9.

Pierron, G. & Sauer, H. W. (1980*b*). More evidence for replication–transcription-coupling in *Physarum polycephalum. J. Cell Sci.* **41**, 105–13.

Polanshek, M. M., Blomquist, J. C., Evans, T. E. & Rusch, H. P. (1978). Amino peptidases of *Physarum polycephalum* during growth and differentiation. *Arch. Biochem. Biophys.* **190**, 261–9.

Prescott, D. M. (1976). *Reproduction of eukaryotic cells.* Academic Press, New York & San Francisco.

Prior, C., Cantor, C. R., Johnson, E. M. & Allfrey, V. G. (1980). Incorporation of exogenous pyrene labeled histone into *Physarum* chromatin, a system for studying changes in nucleosomes assembled *in vivo. Cell*, **20** (3), 597–608.

Rakoczy, L. (1973). The myxomycete *Physarum nudum* as a model organism for photobiological studies. *Ber. Deut. Bot. Ges.* **86**, 141–64.

Rakoczy, L. (1980). Effect of blue light on metabolic processes, development and movement in true slime molds. In *The Blue Light Syndrome*, ed. H. Senger, pp. 570–83. Springer-Verlag, Berlin, Heidelberg & New York.

Raper, K. B. (1935). *Dictyostelium discoideum*, a new species of slime mold from decaying forest leaves. *J. Agric. Res.* **50**, 135–47.

Rapp, P. E. & Berridge, M. J. (1977). Oscillations in calcium–cyclic AMP control loops form the basis of pacemaker activity and other high frequence biological rhythms. *J. Theor. Biol.* **66**, 497–526.

Ron, A. & Prescott, D. M. (1969). The timing of DNA synthesis in *Amoeba proteus. Exp. Cell Res.* **56**, 430–4.

Ross, I. K. (1957). Syngamy and plasmodium formation in the myxogastres. *Am. J. Bot.* **44**, 843–50.

Ross, I. K. (1966). Chromosome numbers in pure and gross cultures of myxomycetes. *Am. J. Bot.* **53**, 712–18.

Rusch, H. P. (1959). The organisation of growth processes. In *Biological Organisation–Cellular and Subcellular*, ed. C. H. Waddington, pp. 263–81. Pergamon Press, New York.

Rusch, H. P. (1969). Some biological events in the growth cycle of *Physarum polycephalum. Fed. Proc.* **28**, 1761–70.

Rusch, H. P. (1970). Some biochemical events in the life cycle of *Physarum polycephalum.* In *Advances in Cell Biology*, ed. D. M. Prescott, L. Goldstein & E. McConkey, vol. 1, pp. 297–327. Appleton-Century-Crofts, New York.

Rusch, H. P. (1980). The search. In *Growth and Differentiation in* Physarum polycephalum, ed. W. F. Dove & H. P. Rusch, pp. 1–6. Princeton University Press, New Jersey.

Rusch, H. P., Sachsenmaier, W., Behrens, K. & Gruter, V. (1966). Synchronization of mitosis by the fusion of the plasmodia of *Physarum polycephalum*. *J. Cell. Biol.* **31**, 204–9.

Sachsenmaier, W. (1976). Control of synchronous nuclear mitosis in *Physarum polycephalum*. In *Molecular Basis of Circadian Rhythms*, ed. J. W. Hastings & H. G. Schweiger, pp. 409–20. Dahlem Konferenzen, Berlin.

Sachsenmaier, W. (1978). Mitose-Zyklen. *Arzneim.-Forsch./Drug. Res.* **28** (II), 10a 1819–24.

Sachsenmaier, W. (ed.) (1979). *Current Research on* Physarum, *Proceedings of the 4th European* Physarum *Workshop*. Publications of the University of Innsbruck 120. Innsbruck University Press.

Sachsenmaier, W. & Ives, D. H. (1965). Periodische Aenderungen der Thymidin-kinase-Aktivität im synchronen Mitosecyclus von *Physarum polycephalum*. *Biochem. Z.* **343**, 399–406.

Sachsenmaier, W., Remy, O. & Plattner-Schobel, R. (1972). Initiation of synchronous mitosis in *Physarum polycephalum*. A model of the control of cell division in eukaryotes. *Exp. Cell Res.* **73**, 41–8.

Sager, R. & Kitchin, R. (1975). Selective silencing of eukaryotic DNA. *Science*, **189**, 426–33.

Sauer, H. W. (1973). Differentiation in *Physarum polycephalum*. *Symp. Soc. Gen. Microbiol.* **23**, 375–406.

Sauer, H. W. (1974). Entwicklungsbiologie, Experimente an Eiern und Pilzen. *Konstanzer Universitätsreden*, vol. 68, ed. G. Hess. Universitätsverlag, Konstanz.

Sauer, H. W. (1978). Regulation of gene expression in the cell cycle of Physarum. In *Cell Cycle Regulation*, ed. J. R. Jeter, I. L. Cameron, G. M. Padilla & A. M. Zimmerman, pp. 149–56. Monographs in Cell Biology Series. Academic Press, New York & London.

Sauer, H. W. (1980). *Entwicklungsbiologie – Ansätze zu einer Synthese*. Springer-Verlag, Heidelberg, Berlin & New York.

Sauer, H. W., Babcock, K. L. & Rusch, H. P. (1969a). Changes in RNA synthesis associated with differentiation (sporulation) in *Physarum polycephalum*. *Biochim. Biophys. Acta*, **195**, 410–21.

Sauer, H. W., Babcock, K. L. & Rusch, H. P. (1969b). High molecular weight phosphorous compound in nucleic acid extracts of the slime mold *Physarum polycephalum*. *J. Bacteriol.* **99**, 650–4.

Sauer, H. W., Babcock, K. L. & Rusch, H. P. (1969c). Sporulation in *Physarum polycephalum*. A model system for studies on differentiation. *Exp. Cell Res.* **57**, 319–27.

Sauer, H. W., Babcock, K. L. & Rusch, H. P. (1970). Changes in nucleic acid and protein synthesis during starvation and spherule formation in *Physarum polycephalum*. *Wilhelm Roux' Arch.* **165**, 110–24.

Sauer, H. W., Goodman, E. M, Babcock, K. L. & Rusch, H. P. (1969). Polyphosphate in the life cycle of *Physarum polycephalum* and its relation to RNA synthesis. *Biochim. Biophys. Acta*, **195**, 401–9.

Sauer, H. W. & Rusch, H. P. (1970). Sporulation bei Physarum; ein Modell genabhängiger Differenzierung. *Zool. Anz. Suppl.* **33**, 350–3.

Schicker, C., Hildebrandt, A. & Sauer, H. W. (1979). RNA transcription of isolated nuclei and chromatin with exogenous RNA polymerases during mitotic cycle and encystment of *Physarum polycephalum*. *Wilhelm Roux' Arch.* **187**, 195–209.

Schiebel, W. (1973). The cell cycle of *Physarum polycephalum*. *Ber. Deut. Bot. Ges.* **86**, 11–38.

Schiebel, W., Chayka, T. A., de Vries, A. & Rusch, H. P. (1969). Decrease of protein synthesis and breakdown of polyribosomes by elevated temperature in *Physarum polycephalum*. *Biochem. Biophys. Res. Commun.* **35**, 338–45.

Schiebel, W. & Schneck, U. (1974). DNA replication continued in isolated nuclei of synchronously growing *Physarum polycephalum*. *Z. Physiol. Chem.* **355**, 1515–25.

Schrauwen, J. (1979). Post fusion incompatibility in *Physarum polycephalum*, the requirement of de-novo synthesized high molecular weight compounds. *Arch. Microbiol.* **122**, 1–8.

Schreckenbach, T., Walckhoff, B. & Verfuerth, C. (1981). Blue light receptor in *Physarum polycephalum* mediates inhibition of spherulation and regulation of glucose metabolism. *Proc. Nat. Acad. Sci. USA*, **78**, 1009–13.

Schultze, B., Kellerer, A. M. & Maurer, W. (1979). Transit times through the cycle phases of jejunal crypt cells of the mouse. *Cell Tissue Kinet.* **12**, 347–59.

Schweinitz, L. D. (1822). 'Synopsis Fungorum'. Naturforschende Ges. in Leipzig.

Seebeck, T. & Braun, R. (1980). Transcription in acellular slime molds. In *Advances in Microbial Physiology*, ed. A. H. Rose, J. G. Morris, vol. 21, pp. 1–45. Academic Press, New York.

Seeman, P. (1972). The membrane actions of anesthetics and tranquilizers. *Pharmacol. Rev.* **24**, 583–623.

Shall, S. (1973). Enzymes of nuclear nicotinamide adenine dinucleotide metabolism. *Biochem. Soc. Trans.* **1**, 648–50.

Shinnick, T. M. & Holt, C. E. (1977). A mutation (gad) linked to mt and affecting asexual plasmodium formation in *Physarum polycephalum*. *J. Bacteriol.* **131**, 247–50.

Shinnick, T. M., Palotta, D. J., Jones-Brown, Y. M., Youngman, P. J. & Holt, C. E. (1978). A gene, *imz*, affecting the pH sensitivity of zygote formation in *Physarum polycephalum*. *Curr. Microbiol.* **1**, 163–6.

Smith, S. S. & Braun, R. (1978). A new method for the purification of RNA polymerase II from the lower eukaryote *Physarum polycephalum*. *Eur. J. Biochem.* **82**, 309–20.

Smith, J. A. & Martin, L. (1974). Regulation of cell proliferation. In *Cell cycle controls*, ed. Padilla, G. M., Cameron, I. L. & Zimmermann, A., pp. 43–60. Academic Press, London & New York.

Spemann, H. & Mangold, H. (1924). Über Induktion von Embryonalanlagen durch Implantation artfremder Organisatoren. *Wilhelm Roux' Arch. Entwicklungsmech. Organismen.* **100**, 599–638.

Strathern, J. N., Spatola, E., McGill, C. & Hicks, J. B. (1980). Structure and organization of transposable mating type cassettes in saccharomyces yeasts. *Proc. Natl. Acad. Sci. USA*, **77**, 2839–43.

Sudbery, P. E. & Grant, W. D. (1975). The control of mitosis in *Physarum polycephalum*: the effect of lowering the DNA : mass ratio by UV irradiation. *Exp. Cell Res.* **95**, 405–15.

Sudbery, P. E. & Grant, W. D. (1976). The control of mitosis in *Physarum polycephalum*. The effect of delaying mitosis and evidence for the operation of the control mechanism in the absence of growth. *J. Cell Sci.* **22**, 59–65.

Sudbery, P., Haugli, K. & Haugli, F. (1978). Enrichment and screening of heat sensitive mutants of *Physarum polycephalum*. *Genet. Res., Camb.* **31**, 1–12.

Sun, J. Y.-C., Johnson, E. M. & Allfrey, V. G. (1979). Initiation of transcription of ribosomal deoxyribonucleic acid sequences in isolated nuclei of *Physarum polycephalum*: studies using nucleoside 5'-[γ-δ]triphosphates and labeled precursors. *Biochemistry*, **8** (21), 4572–80.

Sussman, M. (1966). Protein synthesis and the temporal control of genetic transcription during slime mold development. *Proc. Natl. Acad. Sci. USA*. **55**, 813–18.

Sussman, M. (1976). The genesis of multicellular organization and the control of gene expression in *Dictyostelium discoideum*. *Prog. Mol. Subcell. Biol.* **4**, 103–31.

Taylor, D. L. & Condeelis, J. S. (1979). Cytoplasmic structure and contractility in amoeboid cells. *Int. Rev. Cytol.* **56**, 57–144.

Terayama, K., Ueda, T., Kurihara, K. & Kobatake, Y. (1977). Effect of sugars on salt reception in true slime mold *Physarum polycephalum*: physicochemical interpretation of interaction between salt and sugar receptors. *J. Memb. Biol.* **34**, 369–81.

Truitt, C. L. & Holt, C. E. (1979). Analysis of an Alc mutation. In *Current Research on* Physarum, ed. W. Sachsenmaier, pp. 26–30. Publications of the University of Innsbruck 120. Innsbruck University Press.

Tso, W. & Mansour, T. E. (1975). Thermotaxis in the slime mold *Physarum polycephalum*. *J. Behavior. Biol.* **14**, 499–505.

Tso, W. & Wong, M. (1978). Chemotaxis not a prerequisite for plasmodia coalescence in *Physarum polycephalum*. *Biochem. Biophys. Res. Commun.* **84**, 993–7.

Turner, H. M. & Johnson, J. (1975). A biochemical analysis of induction of microcyst formation in *Physarum polycephalum* myxamoebae by mannitol. *Cytobios*, **13**, 229–39.

Turnock, G. (1980). Patterns of nucleic acid synthesis in *Physarum polycephalum*. *Prog. Nuc. Acid Res. Mol. Biol.* **23**, 53–103.

Tyson, J. J., Garcia-Herdugo, G. & Sachsenmaier, W. (1979). Control of nuclear division in *Physarum polycephalum*. Comparison of cycloheximide pulse treatment, UV irradiation and heat-shock. *Exp. Cell Res.* **119**, 87–98.

Tyson, J. J. & Sachsenmaier, W. (1978). Is nuclear division in *Physarum* controlled by a continuous limit cycle oscillator? *J. Theor. Biol.* **73**, 723–38.

Tyson, J. J. & Sachsenmaier, W. (1979). Derepression as a model for control of the DNA-division cycle in eukaryotes. *J. Theor. Biol.* **79**, 275–80.

Ueda, T. & Kobatake, Y. (1977). Changes in membrane potential, zeta potential and chemotaxis of *Physarum polycephalum* in response to *n*-alcohols, *n*-aldehydes and *n*-fatty acids. *Cytobiologie*, **16**, 16–26.

Ueda, T. & Kobatake, Y. (1978). Discontinuous change in membrane activities of plasmodium of *Physarum polycephalum* caused by temperature variation: effects on chemoreception and amoeboid motility. *Cell Structure and Function*, **3**, 129–39.

Ueda, T. & Kobatake, Y. (1979). Spectral analysis of fluorescence of 8-anilino-1-naphthalenesulfonate in chemoreception with a white plasmodium of *Physarum polycephalum*: evidence for conformational change in chemoreceptive membrane. *Biochim. Biophys. Acta*, **557**, 199–207.

Vandekerckhove, J. & Weber, K. (1978). The amino acid sequence of *Physarum polycephalum* actin. *Nature, London*, **276**, 720–1.

von Stosch, H. A. (1965). Wachstums- und Entwicklungsphysiologie der Myxomyceten. In *Handbuch der Pflanzenphysiologie*, vol. 15, ed. W. Ruhland, pp. 641–79. Springer-Verlag, Berlin, Heidelberg & New York.

von Stosch, H. A., Zyl-Pischinger, M. V. & Dersch, G. (1964). Nuclear phase alternance in the myxomycete *Physarum polycephalum. Abstr. 10th Intern. Bot. Congr. Edinburgh*, pp. 481–2.

Vouk, V. (1910). Untersuchungen über die Bewegung der Plasmodien. Die Rhythmik der Protoplasmaströmung. *Sitzungsb. Kais. Akad. Wiss. Wien*, **119**, 853–76.

Walker, B. W., Christensen, M. E., Beyer, A. L. & LeStourgeon, W. M. (1980). The nuclear proteins of *Physarum polycephalum*: a comparative view. In *Growth and Differentiation in* Physarum polycephalum, ed. W. F. Dove & H. P. Rusch, pp. 98–128. Princeton University Press, New Jersey.

Wallroth, C. F. W. (1833). Flora cryptogamica Germaniae II. Nürnberg.

Waqar, M. A. & Huberman, J. A. (1975). Covalent linkage between RNA and nascent DNA in the slime mold, *Physarum polycephalum. Biochim. Biophys. Acta*, **383**, 410–20.

Ward, J. M. (1959). Biochemical systems involved in differentiation of the fungi. In *Proceedings of the 4th International Congress of Biochemistry, Vienna 1958*, ed. W. J. Nickerson, vol. 6, pp. 33–58. Pergamon Press, Oxford.

Waterborg, J. H. & Kuyper, C. M. A. (1979). Purification of an alkaline nuclease from *Physarum polycephalum. Biochim. Biophys. Acta*, **571**, 359–68.

Weil, A. P., Luse, D. S., Segall, J. & Roeder, R. G. (1979). Selective and accurate initiation of transcription at the Ad2 major late promotor, in a soluble system dependant on purified RNA polymerase II and DNA. *Cell*, **18**, 469–84.

Wendelberger-Schieweg, G. & Hüttermann, A. (1978). Amino-acid pool and protein turnover during differentiation (spherulation) of *Physarum polycephalum. Arch. Microbiol.* **117**, 27–34.

Wheals, A. E., Grant, W. D. & Jockusch, B. M. (1976). Temperature-sensitive mutants of the slime mould *Physarum polycephalum*. I. Mutants of the amoebal phase. *Molec. Gen. Genet.* **149**, 111–14.

Wick, R. (1976). Untersuchungen zur Ribonukleinsäure-Synthese bei *Physarum polycephalum*. PhD Thesis, University of Konstanz.

Wick, R. (1981). Characterization of nuclear DNA from *Physarum polycephalum* extracted from growing, starved and sporulating macroplasmodia. In *Biology of* Physarum, ed. L. Rakoczy, pp. 117–22. Jagiellonian University Press, Krakow.

Wilkins, A. S. & Reynolds, G. (1979). The development of sporulation competence in *Physarum polycephalum. Develop. Biol.* **72**, 175–81.

Wille, J. J., Jr. & Kauffman, S. A. (1975). Premature replication of late S-period DNA regions in early S nuclei transferred to late cytoplasm by fusion in *Physarum polycephalum. Biochim. Biophys. Acta*, **407**, 158–73.

Williams, K. L., Kessin, R. H. & Newell, P. C. (1974). Parasexual genetics in *Dictyostelium discoideum*: mitotic analysis of acriflavin resistance and growth in axenic medium. *J. Gen. Microbiol.* **84**, 59–69.

Wohlfarth-Bottermann, K. E. (1979). Oscillatory contraction activity in *Physarum. J. Exp. Biol.* **81**, 15–32.

Wolf, R., Wick, R. & Sauer, H. W. (1979). Mitosis in *Physarum polycephalum*: analysis of time lapse films and DNA replication of normal and heat-shocked macroplasmodia. *Eur. J. Cell Biol.* **19**, 49–59.

Wormington, W. M., Cho, C. G. & Weaver, R. F. (1975). Sporulation-inducing factor in the slime mould *Physarum polycephalum. Nature, London*, **256**, 413–14.

Wormington, W. M. & Weaver, R. F. (1976). Photoreceptor pigment that induces differentiation in the slime mold *Physarum polycephalum. Proc. Natl. Acad. Sci. USA*, **73**, 3896–9.

Wright, B. E. (1973). *Critical Variables in Differentiation*. Prentice-Hall, Englewood Cliffs.

Wright, M., Mir, L. & Moisand, A. (1979). Ultrastructure of the flagellar apparatus of *Physarum polycephalum* amoeba. In *Current Research on* Physarum, *Proceedings of the 4th European* Physarum *Workshop*, ed. W. Sachsenmaier, pp. 196–201. Publications of the University of Innsbruck 120. Innsbruck University Press.

Wright, M. & Tollon, Y. (1979*a*). *Physarum* thymidine kinase: a step or peak enzyme depending upon temperature of growth. *Eur. J. Biochem.* **96**, 177–81.

Wright, M. & Tollon, Y. (1979*b*). Regulation of thymidine kinase synthesis during the cell cycle of *Physarum* by the heat-sensitive system which triggers mitosis and S-phase. *Exp. Cell Res.* **122**, 273–9.

Youngman, P. J., Adler, P. N., Shinnick, T. M. & Holt, C. E. (1977). An extracellular inducer of asexual plasmodium formation in *Physarum polycephalum*. *Proc. Natl. Acad. Sci., USA*, **74**, 1120–4.

Zaar, K. & Kleinig, H. (1975). Spherulation of *Physarum polycephalum*. I. Ultrastructure. *Cytobiologie*, **10**, 306–28.

Zeuthen, E. & Williams, N. E. (1969). Division-limiting morphogenetic processes in *Tetrahymena*. In *Nucleic Acid Metabolism, Cell Differentiation and Cancer Growth*, ed. E. V. Cowdry & S. Seno, pp. 203–16. Pergamon Press, Oxford.

Author index

Subject Index